Media, Modernity and Dynamic Plants in Early 20th Century German Culture

Critical Plant Studies

PHILOSOPHY, LITERATURE, CULTURE

Series Editor

Michael Marder
(IKERBASQUE / *The University of the Basque Country, Vitoria*)

VOLUME 2

The titles published in this series are listed at *brill.com/cpst*

Media, Modernity and Dynamic Plants in Early 20th Century German Culture

By

Janet Janzen

BRILL
RODOPI

LEIDEN | BOSTON

Cover Illustration: Courtesy of Janet Janzen.

Library of Congress Cataloging-in-Publication Data

Names: Janzen, Janet Lindeblad, author.
Title: Media, modernity, and dynamic plants in early 20th century German culture : / by Janet Janzen.
Description: Leiden ; Boston : Brill Rodopi, 2016. | Series: Critical plant studies: philosophy, literature, culture, ISSN 2213-0659 ; volume 2 | Includes bibliographical references and index.
Identifiers: LCCN 2016027435 (print) | LCCN 2016029938 (ebook) | ISBN 9789004327160 (pbk. : alk. paper) | ISBN 9789004327177 (E-book)
Subjects: LCSH: German literature—20th century—History and criticism. | Plants in literature.
Classification: LCC PT405 .J344 2016 (print) | LCC PT405 (ebook) | DDC 830.9/364—dc23
LC record available at https://lccn.loc.gov/2016027435

Typeface for the Latin, Greek, and Cyrillic scripts: "Brill". See and download: brill.com/brill-typeface.

ISSN 2213-0659
ISBN 978-90-04-32716-0 (paperback)
ISBN 978-90-04-32717-7 (e-book)

This book is printed on acid-free paper and produced in a sustainable manner.

Contents

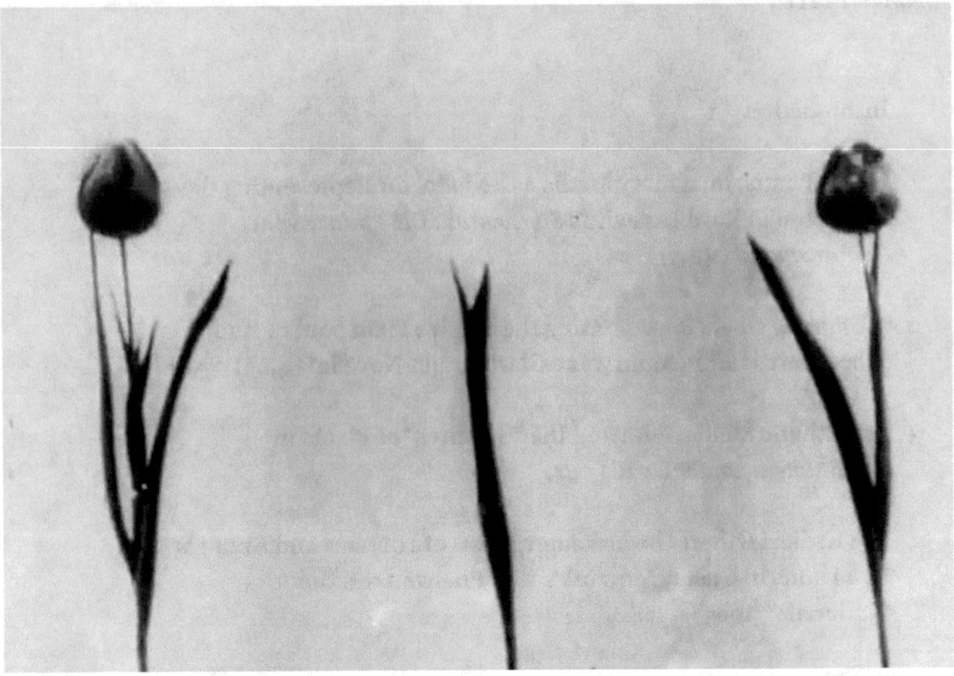

FIGURE 1 *Tulips. Scene still.* Kinematographische Studien an Impatiens, Vicia, Tulipa, Mimosa und Desmodium von W. Pfeffer (1898–1900). *Dir. Wilhelm Pfeffer. 1900.*

Introduction

> In films, we experience through time-lapse photography and slow motion the welling up and down, the breathing and growth of plants. The microscope reveals a world of systems in drops of water, and the instruments of the sky open up the infinitude of space. The technology is what makes our relationship to nature closer than it ever was and opens a window for us into worlds, which were previously closed to our senses. The technology is also what gives us new means to form art. If the highest achievement in art for the century was painting, the phrase, 'The battle of the spirit is being fought out on the canvas,' is justified. Yet, the tools for battle are now iron, concrete, steel... light and airwaves. Our architecture, engineered constructions, cars, planes as much as film, radio, photography are building up proof in a mountain of fantastical possibilities, heights of aesthetic ranges, and thousands of signs that the often bemoaned victory of technology is not a victory of material, but rather of the creative spirit that only manifests itself in new forms.
>
> NIERENDORF, 27

In an early film from 1898, the growth and wilting of several tulips can be seen, waving back and forth against a plain background at the height of their bloom. Labeled simply "Tulips," the film would have been a novel experience for anyone who had happened to see it at the turn of the century while at a film evening or in a travelling exhibition. They would have noticed the rhythmic motions of the accelerated plant movements and some would have gone as far to see an uncanny resemblance to the gestures of animals or even humans. In fact, the purpose of this film was to study the very plant movement that was so novel to early viewers. The film was shot by the German biologist, Wilhelm Pfeffer, who had developed a time-lapse photography apparatus to make the movements of tulips and other plants visible. Time-lapse photography speeds up the movements of plants to the pace of animals and humans by taking photographs at spaced intervals, a method that is still being used today in studies of plant behaviour. Pfeffer photographed plants as a way to counter what he believed was prevailing impression of plants as fixed and non-sentient beings, an impression that Pfeffer also believed was caused by the relatively

 | DOI 10.1163/9789004327177_002

slow pace of plants. By speeding up the tulips' movement, Pfeffer could visualize what he knew to be true, the sentience and dynamism of plants.

At first, these studies of plant movement were limited to the scientific community, but they were introduced to broader public as early as 1900 and became quickly popular, fuelled by a public entranced by the animalistic movements of these plants. The popularization of these scientific studies of movement can be largely attributed to a German film pioneer, Oskar Messter, who played a pivotal role in making these films accessible to the broader public. While Messter is largely known for his film weeklies from the 1920s and his role in early film distribution, he began his in film by producing these time-lapse films of growing plants with the help of Pfeffer's invention, the time-lapse camera. By the 1920s, nature films of moving plants were frequently being shown at film evenings or before main features in front of delighted audiences. The number of films devoted to showing plant movement and sensational plants like the Venus flytrap is astounding, speaking to the wideranging appeal of these plant films. It is clear from Pfeffer's and Messter's contributions that the German science community's role in the popularization of these short time-lapse films is notable. Their contribution to early film history also cannot be discounted, even as they are regularly overshadowed by films that fit into the paradigm of the early 20th century as "against nature."

When we think of German modernism, the first image we recall is not the one of tulips moving at a breakneck speed but a Lumière brothers' film, showing a steam engine rushing towards us and fittingly called *The Arrival of a Train at La Coitat Station* (*L'Arrivée d'un Train en Gare de la Ciotat*). Shot just three years before Pfeffer's tulips, the film repeatedly pops up as an anecdote in scholarship to illustrate the reaction of early film audiences to the ability of film to animate the world through moving pictures. Less overt but also significant is the impression the film leaves of a time preoccupied with technology. The film neatly characterizes European modernism and more specifically German modernism as an age, where technology animated the world and shaped the urban landscape. The other image of modernism, Pfeffer's time-lapse film of growing tulips from 1898, receives less attention for it offers a more nuanced view of German modernism—one that does not see technology pitted against nature, but at times bringing people closer to nature. As seen in Pfeffer's studies of plant movement, technology does play a significant role and this should not be diminished. Yet, the emphasis can be shifted to also acknowledge the significant role of nature in modernism—a project that has just started to be realized by research such as Axel Goodbody's *Natur, Technology and Cultural Change in Twentieth-Century German Literature*, and Oliver Botar's and Isabel Wünsche's *Biocentrism and Modernism*. It is the project of this book to continue this re-examination by reintegrating nature into our understanding of German modernism, using plants as a formative example.

Pfeffer's studies in plant movement characterize one of the key strands of this book, how new media such as film helped to fundamentally alter our relationship to nature as well as the way we qualify and categorize other beings. His movement studies and other early films seemed to reveal a previously unknown set of commonalities between people, animals and plants, which persisted in the subsequent nature films and in literary texts. Thirty years after Pfeffer's plant films, a feature-length film, *Das Blumenwunder* (*The Miracle of Flowers*, 1926), directly illustrates the effect time-lapse films had on the representation and understanding of plant life. Discussed in chapter three, the film suggests through time-lapse images of growing plants that plants are alive like us, and then attempts to find a means of understanding plant life. However, it would be incorrect to state that the view of plants as sentient and dynamic beings in the early 20th century was simply a product of new media. Rather, the time-lapse images helped to provide evidence for an existing conception of plant life that was already forming in the late 19th century in reaction to the proliferation of materialistic, positivist and naturalistic tendencies. The topic of chapter one, Kurd Lasswitz's novel *Sternentau: Die Pflanze vom Neptunsmond* (*Star-dew: The Plant from Neptune's Moon*, 1909) represents one of the earliest examples of this reaction, combining the empirical practice of science with pan-psychism to explore the possibility of plant sentience and communication. In a short story by Paul Scheerbart, discussed in the second chapter, plants are represented as manifestations of nature's creative dynamic force as opposed to the reductionist perspective of materialism and naturalism. He derives this perspective on plants from Romanticism, and from the Vitalist currents of science and philosophy that grew out of Romanticism in the 19th century.

The capacity to see plant movement did not always lead to a shift in the perception of plants. Rather, alongside the positive and even utopian reception of plant life as mobile and sentient beings, a more conservative perspective persists. It maintains the definitions and distinctions of categories of being as well as an instrumental view of nature, the belief that the purpose of plants is to be eaten and enjoyed by humans. From this point of view, plants that appeared to pose a challenge to traditional distinctions of living beings, such as *Mimosa pudica* (which reacts instantly to contact) or the insect-eating Venus flytrap, took on a particularly charged status as monstrous anomalies of nature. Films like *Nosferatu* (1922) and *Alraune* (1928), and short stories like Gustav Meyrink's "Die Pflanzen des Doktor Cinderella" ("The Plants of Dr. Cinderella," 1905) associate such plants (the Venus flytrap) with characters in the position of the radically other. In so doing, they reveal such conservative tendencies to be social anxieties. The manner in which plants are used as a model for an upheaval in social hierarchies, as well as a reflection on how the short story and the two films reinforce or criticize this instrumentalization of plants, will be the focus of the last two chapters.

Within the selection of texts I chose for this book, there is a spectrum of reactions that characterizes our perception of and relationship to plants as, on one end, one that welcomes commonalities between plants, animals and people, and on the other end, those that do not. This perception of and relationship to plants is the focus of a recent trend in scholarship and mainstream journalism that seeks to examine and reconfigure it.[1] At the center of the discussion are questions concerning plant communication, sentience and even ethical concerns. Philosopher Matthew Hall examines the perception and role of plants from ancient Greece to the present day in his book *Plants as Persons* (2011). Since ancient Greece, the plant's apparent lack of movement and sentience allowed for many philosophers and botanists to argue that they represented a form of life categorically inferior to animals and humans, giving way to a persisting anthropocentric and zoocentric perspective. As Hall argues, plants are often placed below humans and animals in a hierarchy of being because of their position as radically other. Their multivariate form, immobility, the missing nervous system and the lack of an apparent seat of the human mind (the brain) have been used to focus on differences rather than affinities between humans and plants and to absolve any moral qualms that would come from the necessary killing of plants for food. Instead of attempting to perceive the plant on its own terms, plants are viewed instrumentally, as existing for the purpose of serving the needs of humans. In literary and visual texts, the instrumental perspective of plants is apparent in their function in the narrative as symbols for human experience rather than for nature.

Like Hall, Elaine Miller writes from within the philosophical tradition, but instead of attempting to give plants personhood, she is working within the human hierarchy using plants to propose an alternative feminine subjectivity and society. Her book, *The Vegetative Soul: From the Philosophy of Nature to Subjectivity in the Feminine* (2002), draws on the German philosophical

1 For other scholars and works not discussed here, Randy Laist ed., *Plants and Literature: Essays in Critical Plant Studies*, (Amsterdam: Rodopi, 2013); D. J. Beerling, *The Emerald Planet: How Plants Changed Earth's History* (Oxford: Oxford University Press, 2007); and Daniel Chamovitz, *What a Plant Knows: A Field Guide to the Sense*, (New York: Scientific American/ Farrar, Straus and Giroux, 2012).

tradition, from Immanuel Kant to Johann Wolfgang von Goethe, Friedrich Hölderlin and Friedrich Nietzsche, to lay out an alternative plant subjectivity as a model for human subjectivity. The vegetative soul, as she calls it, emphasizes interdependence and metamorphosis over a human subjectivity's focus on individuation and atomism. While she does not attempt to understand a plant on its own terms, her approach implicitly values the plant mode of being as an answer to the inequality in the hierarchy of being perpetrated by the anthropocentric and zoocentric world views. Miller ends with a discussion of Luce Irigaray's use of plant imagery to open up the possibility of a "plant-like" reading of texts as fluid entities.

Michael Marder's *Plant Thinking: A Philosophy of Vegetal Life* (2013) is the latest contribution to the small but significant turn to plant studies within philosophy. Like Hall and Miller, Marder is seeking to give "new prominence to vegetal life" and to examine the assumptions that structure our understanding of life (3). This book shares with Miller's the fundamental question concerning the encounters between plants and humans. How can revisiting our representations of plant-human encounters change the way we interact with plants in our daily lives? The interactions between humans and plants have been often overlooked in culture studies in the same fashion that plants often serve as the background to our lives. However, contributions such as *Green Sense: The Aesthetics of Plants, Place and Language* (2012) by John Ryan, serve as one of the ground breaking steps to creating a plant studies field within cultural studies. In the short stories and films that are discussed in this book, the perspective of plants as dynamic entities either proves to be the hinge that opens up an alternative social structure and mode of being or it reveals the anxieties associated with its radical character.

The unusual prose and method of another philosopher, Michel Serres, seems to exemplify a plant-like cultural interpretation as set out by Matthew Hall. Michel Serres' book, *The Natural Contract* (1995), has in common with Hall's *Plants as Persons* the conviction that considering the natural world as a person would bring about change in our relationship with "her." Instead of focussing his argument on plants as Hall does, Serres suggests that nature's perspective as a whole should be considered when rethinking the social contract. After tracing the intertwined history of science and law, his book proposes that we are no longer in a position to consider nature merely as a threat to our survival or as an object to be dominated but should treat nature rather as an equal partner. As he argues, the threat of global warming is in effect the voice of nature demanding a part in the renegotiation as well as recognition of the often-overlooked connection between humans and the environment. He describes the connection in terms of the familiar metaphor of mother nature: "In arm wrestling, with an umbilical cord, in the sexual bond? All that and more. The cords that tie us together form, in all, a third kind of world: they are nutritive, material, scientific and technological, informational, aesthetic, religious. Equipotent to the Earth, we have become its biplanet, and it likewise becoming our biplanet both bound by an entire planet of relations" (110). Serres concludes by challenging our capacity to raise the status of the Earth to equal partners: "Would I acknowledge her as my mother, my daughter, and my lover together? Should I let her sign?" (124).

Within pop culture, numerous films and books have surfaced asking if plants have intelligence. Michael Pollan's book, *The Botany of Desire* (2001), takes the position that plants have agency and manipulate human desires to propagate their species, spreading across the Earth. He focuses on four plants, which he views as manipulating four human desires: tulips for beauty, marijuana for pleasure, potatoes for sustenance and apples for sweetness. His inversion of the perspective of plants as passive and subject to our manipulations has gained a great deal of attention, so much so that PBS found reason to release a documentary based on his ideas under the same title. The popularity of Pollan's ideas is partly due to his unusual stance on the plant-human relationship and the sensationalism that often accompanies any discussion of human desire. However, Pollan does tap into the persisting perception of plants as immobile and passive, and the crux of his argument rests on seeing plants as mobile and active.

Pollan's provocative view of plant agency is balanced by the careful consideration of plant communication and sensing by the scientist Richard Karbin. His book *Plant Sensing and Communication* (2015) is a key contribution to the growing field of plant neurobiology, which proposes that the interaction of plants with their environment may signal behaviour and even intelligence. Far from the anthropomorphic speculations of Peter Tompkins' *The Secret Life of Plants* (1973), Karbin avoids projecting human emotion and intelligence on plants and instead concentrates on the biological basis for plant sensing and communication on the cellular and molecular level. Like Karbin, the scientist Anthony Trewavas' *Plant Behaviour and Intelligence* (2015) limits his assessment of plant intelligence and behaviour to the physiological and anatomical constraints of plants. Plant intelligence is defined as both the ability to respond to changes in the environment and to compete for resources.

These new discoveries in the natural sciences that highlight plant mobility and sentience have also made their way into the mainstream media. One example comes from a Canadian documentary, "Smart Plants: Uncovering the Secret World of Plant Behaviour," produced as a part of David Suzuki's series, *The Nature of Things*. The show uses easy-to-understand analogies to visualize recent discoveries from biology in the field of plant behaviour, by drawing comparisons between the foraging behaviour of animals and the seeking behaviour of roots. Plant sentience, the ability to sense environment and react to it, is another major focus of the documentary, illustrated through the example of the reaction of grass to being cut by a lawnmower. In response to being cut, the grass releases a chemical that attracts predatory insects like mosquitoes and, with luck, drives away whatever is damaging the grass. David Suzuki's film belongs to a recent renewed interest in plant intelligence in television and film. In all of these films and books,[2] new interpretations of the growth of plants as well as new evidence on a molecular level from the field of biology gives, at the very least, cause to re-evaluate our relationship to plants as well as our preconceived ideas of the life of a plant.

Examining how the relationship of plants and humans has been represented in modern literature and film has the potential to contribute to a larger field of study concerned with our impact on the environment, which has become more urgent in recent years as evidence mounts on the side of global warming. While environmentalism in literature is not a focus of this book, inevitably the theme of human destruction of plants crops up when looking at representations of plants in film and literary texts. As the recent interest in the relationship of plants and humans shows, understanding the cultural significance of plant representations reveals underlying assumptions that drive our behaviour in relation to plants. The films and short stories discussed in this book reveal the complexity in our relationship to plants. On the one hand, there is a desire to have a relationship to nature based on mutual creativity, dynamism, and communication. On the other hand, plants are used as metaphors for anxieties about gender, nationality, categories of being and innovations in science, anxieties that reveal resistance to change and to fluid identities. Innovations that changed the way we see and our relationship to nature are in part responsible for reviving an old debate between a static and rigidly differentiated world and a dynamic and intermingled world.

The debate concerning the sentience of plants can be traced back to the differing perspectives of Aristotle and his pupil Theophrastus. Aristotle attributed a limited awareness to plants in compared to capacities he attributed to animals and humans. Plants, according to Aristotle, possess the most inferior degree of soul, the "nutritive" soul, while animals have both the "nutritive" and "perceptive" soul, and humans have all three, the last being the "rational" soul. Plants' interaction with the world was limited to seeking nutrients placing

2 Other films include: the PBS documentary, *The Plant* (1994); the BBC documentary series *The Private Life of Plants* (1995) and *Kingdom of Plants* (2012) just to name a few.

them below animals and humans in this early hierarchy of being. Aristotle's pupil Theophrastus, who wrote more extensively on plants than his teacher, leaves a more ambivalent and contentious legacy.

The disagreement arises on the extent to which the "father of botany" Theophrastus departed from Aristotle's view of plants as inferior and insentient. From the fragments of his work and what has been quoted by others, as Hall argues, it appears that Theophrastus was not only open to the possibility that plants were sentient beings, capable of responding to their environment but also he considered plants to be on par with humans and other animals. Hans Ingensiep holds a very different view in his *Geschichte der Pflanzenseele* (*History of the Plant Soul*, 2001), and argues that Theophrastus' view of plants is congruent with Aristotle's, excluding plants from the possibility of sentience. Even if Theophrastus believed in plant sentience, this aspect of his legacy has been marginalized by modern society, and by Aristotle's tripartite soul continues to influence philosophy and mainstream culture. After the Enlightenment, however, Aristotle's ontology of life-forms did not go unchallenged.

In German culture, one of the more significant challenges to Aristotle's view of plants came from Johann Wolfgang von Goethe's exploration of plant life. Goethe developed the idea of a plant subjectivity from his empathetic observations of plant life. Goethe's method of observation inherently prevents the view that plants exist solely for the purpose of human use by attempting to understand plants from within. He did recognize that any understanding of nature would be shaped by the person observing, yet still believed that it was possible to think "objectively." Objective here does not refer to the current sense of the word—meaning free from subjective interventions between nature and representation (Daston, The Image of Objectivity, 82)—but rather to the attempt to combine observation with an awareness of the theory that colours all scientists' conclusions and be, therefore, suitably cautious.[3] This method of observation would enable a move from observation to "higher principles of connection" and lead to a "kind of cognition that might constitute an adequate idea and ultimately an intuition into the whole [...]" (Richards 439). As Miller has shown, Goethe came to the conclusion through his objective method that plants and nature as a whole can be understood in terms of rhythm, which consists of contraction, expansion, and intensification (29).[4] Goethe called the three forces combined, metamorphosis.

---

3 See also Lorraine Daston and Peter Gaston, Objectivity, (New York: Zone Books, 2007), for a thorough examination of the historical shifts in meaning of the term "objective."

4 Elaine Miller argues that Goethe is referring to the Greek meaning of rhythm, when he describes plant movement as such (29). Rhythm in this sense means an instance of flow (ρέω).

For Goethe, plants illustrate metamorphosis in nature with great clarity. This is what he meant by the famous statement "Alles ist Blatt!" ("Everything is Leaf"). Every part of the plant, as he considers, either turns into leaf or has just transformed out of leaf, which occurs through contraction, expansion or intensification. Plants are, according to Goethe, in constant movement and are virtually indistinguishable from their offspring, which prevents them from being considered solely in terms of their form and as individuals. Such individuation was precisely the mistake Goethe found in Linnaeus' taxonomic system of classification. The differing approaches of Linnaeus and Goethe to the study of plants illustrate a divide in the botanical sciences throughout the 19th century. Goethe's perception of plants as plural and dynamic, and his method of observation, initiated the concept of the vegetative soul in the tradition of German letters and philosophy.

The Goethean perception of plants as a system of dynamic forces influenced an entire generation of romantic writers including Friedrich von Hardenberg (Novalis), Friedrich Hölderlin, Friedrich Wilhelm Schelling, Friedrich Schlegel,[5] and August Schlegel among others, many of whom applied the plant model to human subjectivity. Briefly and by no means comprehensively, these writers and philosophers believed that knowledge of nature was necessary for knowledge of the self and often incorporated their knowledge of botany and other sciences into their writing. They were reacting to the vocabulary and ideas of the previous generation by moving away from the perception of nature as a series of discrete and classifiable objects to a view of nature as a process. This concept of nature helped to shape a shift during 19th century in the perception of nature to that of a dynamic process. Certainly, Schelling's choice of words readily leads to this interpretation, describing nature as "active" (8), "always changeable" and in "continuous evolution" (12).[6]

This dynamic view of nature appears in the use of plant metaphors by many romantic writers. The plant captures the entanglement between the person and the environment in the close relationship between two life forces working on the plant, the open-ended growth from within and the nourishment

---

5 Friedrich Schlegel's Lucinde (1799) justifies the need for idleness to create art and obtain knowledge through a comparison with the passivity of the plants and its place as the most beautiful and cultivated forms in nature. The most accomplished life is called "ein reines Vegetieren" ("a pure vegetating," 47).

6 Schelling's descriptions of nature as dynamic were taken from *Erster Entwurf eines Systems der Naturphilosophie*: "Wir kennen die Natur nur als thätig" (8). "...kein permanentes Daseyn" (12). "...jedes Product, das jetzt in der Natur fixiert erscheint, würde nur einen Moment existiren, und in continuirlicher Evolution begriffen, stets wandelbar, nur erscheinend vorüberschwinden" (12).

from the environment. In her interpretation of Friedrich Hölderlin's *Hyperion*, Elaine Miller sees "the metamorphosis of plants as a figuration of human life itself" (66). The character Hyperion's development follows the rhythm of contraction and expansion towards a final completion that will never be realized. Key to Miller's interpretation is that the subject can never be the same person he was before.

The significance of plants for the romantic concept of self is perhaps best encapsulated by what has become the symbol of German romanticism, the blue flower. Introduced by Novalis' fragment novel *Heinrich von Ofterdingen* (1802), the blue flower illustrates the role that nature would come to play as itself a work of art, expressive of a deeper unity with humans as a part of the natural world rather than distinct from it. Similarly, as Friedrich Schlegel in 1799, "the world as a whole and originally, is a plant," meaning that the natural world and the function of art insofar is an unfolding, dynamic whole that remains open and incomplete.

After the brief flowering of romanticism in the early 19th century, the ideas of Goethe and his disciples were marginalized by the rationalist science of the industrial era. Scientists like Gustav Fechner were the exception.[7] Fechner found himself asking philosophical questions about the nature of plant life after experiencing a devastating health issue (Heidelberger 54). Inspired by Goethe as well as his own experience with extraordinary plants such as the *Mimosa pudica*, Fechner argued for plant sentience in his book on plants, *Nanna, oder über die Intelligenz der Pflanze* (*Nanna, or Concerning the Intelligence of the Plant*, 1848). He challenged the view that plants are passive, receptive and immobile, and questioned the taxonomies that distinguish lower level animals from higher level plants. Fechner confronts the two main objections to plant sentience, the lack of a brain as a seat for the mind and the difficulty in ascertaining any level of intentional movement similar to humans and animals. With plants such as the *Mimosa pudica* that visibly respond to their environment, it is easy for Fechner to conclude that plants have a sensory-rich inner life formed by their relationship to their environment and their own

7 In his book on Gustav Fechner, Michael Heidelberger details the across-the-board rejection Gustav Fechner experienced from the scientific community in response to his idea of plant and cosmic souls. The same rejection was matched by acceptance by women readers. Just after Nanna was published, one of Fechner's readers sent him a crooked carrot in the mail as a token of her appreciation of his ideas. The carrot came to represent for him the reception of his work (57).

purpose (201).[8] In an insight later echoed by Michael Hall, Fechner sees the last barriers to recognizing a plant's intelligence as lying not in the plant world but with ourselves. Believing that a plant could be a sentient being, which is trampled underfoot and lives to serve the plates and needs of beings higher on the food chain, offends the humane person's sensibilities.

By reframing plants as sentient beings, it follows that Fechner would question the authority of the botanical taxonomy and the corresponding hierarchy of being that places all plants below even the simplest of animals. According to Fechner, the taxonomy creates an artificial hierarchy of being that does not fully correlate to the level of complexity of plants and various animals. That taxonomy differentiates higher plant forms from lower animal forms—for example the *Adansonia digitata* tree (African baobab, *Affenbrotbaum*), which can live as long as a thousand years, is placed below simple animals like the hydra with its branch-like appearance, and microorganisms (*Infusionstierchen*) (252). However, even as the complexity of the *Adansonia* tree fascinates Fechner and causes him to question our understanding of the hierarchy of being, he refrains from eliminating the hierarchy of being completely, as is clear from his comparison of plants with women and children.[9] Fechner places

8 In Fechner's words: "Denn das größte Wunder der Natur liegt doch darin, daß jedes ihrer Wesen in jedem Bezirke, indem es ganz für andere Wesen gemacht erscheint, zugleich ganz auf eigne Zwecke gestellt bleibt, eins immer dem andern dient, nach andrer Beziehung nur, als Andres ihm wieder dient; und Alles dabei so abgewogen in einander greift, daß das Ganze haltbar und lebendig besteht" ("For the biggest miracle of nature can be seen in the fact that everyone of her creatures in every area remains at the same time focussed on its own purpose even though it appears made for another—one always serves another while in another relationship another one serves always him, and because of that everything is so intertwined with another that the whole is lasting and lively") (201).

9 His comparison of women, children and plants tells a great deal of the assumed role of women and children in the mid-nineteenth century in addition to the hierarchy of men over women and children: "Die Pflanze bleibt, wie das Weib dem Manne, immer dem Willen des Thieres unterthan, kommt ihm aber selbst im schönsten Verhältnis, wie es der Schmetterling zur Blume zeigt, nicht entgegen. Sie plaudert gern duftend mit ihren Nachbarinnen. Sie sorgt für die Nahrung des Thieres, bäckt Brod (in den Aehren), bereitet Gemüse für dasselbe. Ihr liebstes Geschäft aber bleibt bis zur Blütezeit ihres Lebens, sich schön zu schmücken und ihrer Gestalt immer neu und schön darzustellen" ("The plant remains always subordinate to the will of the animal like the woman to the man. This fact does not contradict the beautiful relationship they have, for example the butterfly and the flower. They like to chat with their neighbours, take care of the nourishment of the animals, bake bread (in the air), prepare vegetables for the same. Their favourite business remains, however, at the flowering of their life, to make themselves up beautifully and to display their shape in constantly new and beautiful ways") (349).

plants underneath animals, bound to the will of animals, like women and children are bound to men, very much in line with the understanding of the role and capacity of women and children in the Biedermeier culture of his day from 1815 to 1848. The plant soul remains in the developmental stage of children, caught up in their sensuous nature and in a relationship with mother earth as children are to their mothers. Fechner's most radical idea appears still to be his argument for a plant soul, which in his later work *Zend Avesta* (1851) comes to encompass the whole of nature in his concept of a cosmic soul.[10]

Fifty years after *Nanna* had first been published, Fechner's philosophical world view begins to find the respect and acclaim missing at the time. His ideas resonated with writers who were reacting to the representation of the natural world by proponents of naturalism and materialism by seeking out an alternative to analytic taxonomies. They found this alternative in Fechner's concept of a plant subjectivity, dynamic and interdependent. As the introduction to the second edition of *Nanna* in 1908, written by a forerunner to science fiction Kurd Lasswitz, attests, the world was now ready for Fechner's view of nature as a conscious entity (IV 1908):

> But a philosopher took up the thought [of a plant soul] and examined with care, whether there is some truth to the fairy tale. Whoever would

"[...] daß die Frauen selbst doch immer nur Kinder gegen die Männer bleiben" ("[...] that the women themselves remain only children in comparison to men") (347).

"Die Vergleichspuncte der Blumen mit Kindern liegen darin, daß sie, die Erde als ihre gemeinschaftliche Mutter betrachtet, noch an ihr hängen, aus ihr die Nahrung saugen; daß sie sich alle Bedürfnisse zubringen lassen; nicht in's Weite laufen; daß sie lieblich, freundlich, unschuldig aussehen; niemand etwas zu Leide thun; helle Kleider anhaben, und, wie wir meinen, mit ihrer Seele noch eben so im Sinnlichen befangen sind, als es die Kinderseele ist" ("The points of comparison of the flowers with children can be observed in their relationship to the Earth as their common mother. They are still connected to her, suck nourishment from her; all needs are being taken care of by her. They don't run free in the broad expanse. The plants appear sweet, friendly, innocent; never do anyone harm, wear bright clothes, and, in our opinion, are caught in the sensory world with their soul as if it were a child's soul," 347).

10 In the introduction to *Zend-Avesta*, Fechner lists *Nanna* as the precursor to this book: "Eine frühere Schrift, Nanna, kann insofern als Vorläuferin der jetztigen gelten, als dort wie hier versucht wird, das Gebiet der individuellen Beseelung über die gewöhnlich angenommenen Gränzen hinaus zu erweitern; dort aber in abwärts gehender, hier in aufwärts gehender Richtung" ("An earlier text, Nanna, can be considered the predecessor to this one, since here as much as there an attempt is being made to expand the usual assumed borders of the area of the individual ensoulment—there in a downward progression and here and an upwards direction," IV).

> like to know more, should read this book about the soul life of plants, which a fine and sharp spirit wrote and published for the first time in 1848. At that time, many in the educated public shook their heads over it, and fifty years is a long time to wait for a second printing. But the fact that a new edition is at all necessary after fifty years is a good sign that it was a good book—a book that has lasting meaning in and of itself as much as because of its writer. And now, when it is being read in constantly new editions, those in the educated public are not dismissing it as much as before. For the times have changed. (IV)[11]

Lasswitz went on write a novel and several short stories and essays, critically exploring Fecher's concept of plant conscious. As will be discussed in chapter one, Fechner's argument for plant sentience is viewed critically through Lasswitz's background in science and philosophy. Plants are not just represented as dynamic, sentient beings, but the potential to communicate with plants is being explored.

Fechner's cosmology in particular also spoke to Lasswitz's contemporary Paul Scheerbart, whose fiction and many articles were often inspired and influenced by Fechner's vision of a plant soul. As will be discussed in the second chapter of this book, the natural world and particularly plants were represented in Scheerbart's short story, "Flora Mohr," as dynamic entities growing and changing and in close relationship with the cosmos—ideas that came from Fechner and also in part from Goethe. Another factor may have influenced the reception of Fechner's ideas just fifty years later, when Scheerbart was writing. Advances in film along with time-lapse photography by pioneers like Wilhelm Pfeffer played a large role in contributing to the visibility and believability of organic and inorganic nature in motion.

Scheerbart and Lasswitz were not alone in their attraction to Fechner's dynamic vision of the universe at the turn of the century. I would like to highlight the work of two authors, the symbolist writer Maurice Maeterlinck and

---

11 "Aber ein Philosoph hat den Gedanken aufgegriffen und mit Sorgfalt untersucht, wie viel hinter dem Märchen Wahrheit stecke. Wer es genauer wissen will, der lese dieses Buch vom Seelenleben der Pflanze, das ein feiner und scharfer Geist schuf und im Jahre 1848 zum ersten Male herausgab. Man hat damals in der gelehrten Welt viel den Kopf darüber geschüttelt, und fünfzig Jahre bis zur zweiten Auflage ist eine lange Zeit. Aber daß überhaupt nach fünfzig Jahren eine neue Auflage nötig wurde, das ist ein sicheres Zeichen, daß es ein gutes Buch war, ein Buch, das eine Bedeutung hat für Dauer, sowohl durch sich selbst wie durch seinen Verfasser. Und wenn es jetzt in immer neuen Auflagen wieder gelesen wird, so ist auch des Kopfschüttelns in der gelehrten Welt viel weniger geworden. Denn die Zeiten haben sich geändert."

the naturalist Raoul Heinrich Francé, to illustrate the influence of Fechner's interpretation of plant movement at the turn of the century.[12] The first of the two, Maurice Maeterlinck, emphasizes the connection between movement and a kind of plant intelligence in two botanical essays, "The Intelligence of Flowers" and "Scent," both published in 1907. While he is largely known for his symbolist plays, the two essays showcase his considerable knowledge of botany. Drawing from examples in the natural world, Maeterlinck argues in poetic prose that a plant's destiny, to be rooted in one spot and apparently immobile, is overcome by inventive movement that reveals its intent and initiative. In his words, a "revolt against destiny" is hiding underneath the plant's apparent passivity (2). These conclusions are catalyzed by Maeterlinck's own observations of a hundred-year-old laurel tree's compensation for a precarious position on a cliff face. By bending its trunk as it grows and later reinforcing its position with two roots at a particularly appropriate moment, the distressed tree appears to foresee a fall to its demise and plans to prevent it. In a caveat to his interpretation of plant behaviour, Maeterlinck also wonders if the tree is saved by mere coincidence (Maeterlinck 9).[13] On a more fundamental level, movement also surfaces in Maeterlinck's descriptions of the reproductive process as integral to a plant's success. His descriptions of plant reproduction are filled with a language of motion, using words like "scattering," "propulsion," and "aviation" (4). The picture of dynamic nature in Maeterlinck's nature writings is all the more meaningful when viewed as a part of his project to communicate the latest research in botany to the public. Furthermore, his research reflects the increasing acceptance of Fechner's insights into plant movement under the influence of a world that has suddenly become animated—through

12 Others interested in plant intelligence and Fechner from the turn of the twentieth century include: Heino Hayungs' *Die Lehre von der Beseeltheit der Pflanze von Fechner bis zur Gegenwart* (1912), whose dissertation focussed specifically on soul and plants; Erich Becker's *Deutsche Philosophen: Lebensgang u. Lehrgebäude von Kant, Schelling, Fechner, Lotze, Lange, Erdmann, Mach, Stumpf, Bäumker, Eucken, Siegfried Becher* (1929), and Herbert Brunzlow's *Über die Anwendung psychologischer Kategorien auf Pflanzen bei Fechner und Francé: eine historisch-logische Studie* (1920).

13 Here is Maeterlinck's complete description of the tree that also includes a reference to the relative slowness of plant movement: "Then, obeying goodness knows what order of the instinct, two solid roots, two hair cables, emerging from the trunk at more than two feet above the bend, came to moor it to the granite face. Had they truly been brought forth by distress, or else had they been waiting, perhaps with foresight, since the first days, for the critical hour of danger in order to enhance the value of their assistance? Or was it just a happy coincidence? What human eye will ever capture these silent dramas, too long-lasting for our brief lives?" (Maeterlinck 9).

the increasing popularity of moving pictures and through inventions that were increasingly present in daily life: the streetcar, the automobile and even electricity among others.

A push to accept plants as dynamic and sentient can also be found in Raoul Heinrich Francé's many writings.[14] Like Fechner and Maeterlinck, Francé was preoccupied with the idea of a plant intelligence comparable to that of animals and visible through movement. In his book, *Das Sinnesleben der Pflanze* (*The Germ of Mind in Plants*, 1905), published just two years before Maeterlinck's essay from 1907, Francé lays out his objections to a mechanized view of life while simultaneously arguing for an animated nature. For Francé, the key to understanding human origins, our place in the world and our psyche could be found in understanding plants as dynamic, living beings, which participate in relationships with others in their environment through movement.[15] The following quote from *Das Sinnesleben der Pflanze* is typical of his work and the themes he addresses, but also reveals his shared concerns with writers and filmmakers of the 1920s and 30s:

> But the plant also moves its whole body so freely, easily and gracefully as the most skillful animal—only slower. The roots burrow searchingly in the earth. The buds and shoots complete measured circles. The leaves and blossoms nod and shiver by changes. The vines circle searchingly and reach with a ghostly arm into the surroundings. But the superficial

---

14 Raoul Francé's writings are numerous and include other books which discuss plants as sentient beings: Pflanzenpsychologie als Arbeitshypthese der Pflanzenphysiologie (Plant Psychology as a Working Hypothesis of the Plant Physiology, 1909), Die Pflanze als Erfinder (The Plant as Inventor, 1920), and Das Liebesleben der Pflanzen (The Love Life of Plants, 1919). Writing slightly before Francé, Wilhelm Bölsche was also a great popularizer of scientific themes. He is most famous for his seminal essay "Die Naturwissenschaftlichen Grundlagen der Poesie" (The Scientific Basis for Poetry," 1887), yet it is his popular science treatise on origins of love Das Liebesleben in der Natur (Love Life in Nature, 1896) that bears some resemblence to Francé book on the love life of plants.

15 Francé explicitly compares humans and plants: "Es ist etwas Ahnliches in den Pflanzen wie in unserer eigenen Brust" ("There is something similar in plants as in our own chest," 10). Due to the similarities, he draws the conclusion that we can learn about ourselves through learning about plants: "Vorbei war die Zeit der geistlosen Blatt- und Blütenbeschreibungen, ein neues Leben war auch in der Botanik erwachsen und in der letzten Generation wurde sie etwas ganz anderes, eine Fortsetzung oder wenn man will, der Anfang der Erkenntnis von der wahren Natur des Menschen" ("The times of the spiritless descriptions of leaves and blossoms are gone. A new life is also arising in botany and in the last generation, it is becoming something completely different—a continuation or in other words—the start to a new knowledge of the true nature of humans," 11).

> person goes by and considers the plant to be fixed and lifeless, because he does not take the time to dwell for an hour at its side. The plant has the time—that's why it doesn't hurry. For the giants in Flora's realm live through centuries and see at their feet the countless generations of people living and dying. (14).[16]

Movement permeates Francé's description of the plant linking rather than distinguishing it from animals and humans. The words he chooses describe plant movement as gestures, implicitly suggesting that plants communicate and interact with their environment in a manner similar to animals and humans. Plants only appear to be immobile and unaware of their environment from the slow pace of their movements, making their time relatively drawn out in comparison to humans.

Films of plants in time-lapse address this discrepancy in pace between humans and plants, which leads to misconceptions on the nature of plant life and their mobility. Films were particularly appropriate for illustrating Francé's animated plant world, a point acknowledged by Francé and then by a film adaptation of this book *The Germ of Life in Plants* (86).[17] If Francé is asking here that plants be seen as dynamic, sentient beings, he is also attempting to revise the relationship between plants and humans to a more egalitarian one.

Another theme—or in this case controversy—common to many of the texts discussed in this book is visible in Francé's writings as a debate between taxonomy and ecology. At the time of Francé's writing, ecology was a new term coined by Ernst Haeckel[18] and applied to plant communities by the Danish

16 "Aber die Pflanze bewegt auch ihren ganzen Körper so frei und leicht und graziös wie das geschickteste Tier—nur viel langsamer. Die Wurzeln wühlen suchend im Erdreich, die Knospen und Sprosse vollführen gemessene Kreise, die Blätter und Blüten nicken und schauern bei Veränderungen, die Ranken kreisen suchend und langen mit gespenstigem Arm nach der Umgebung—aber die oberflächliche Mensch geht vorbei und hält die Pflanze für starr und leblos, weil er sich nicht die Zeit nimmt, eine Stunde lang bei ihr zu weilen. Die Pflanze aber hat Zeit, darum eilt sie nicht; denn die Riesen in Floras Reich leben durch die Jahrtausende und sehen zu ihren Füßen ungezählte Generationen von Menschen aufleben und vergehen."

17 One of Francé's books *Das Sinnesleben der Pflanzen* (1905) was adapted to film more than 30 years later by a prolific director of nature films and a botanist, Dr. Ulrich Schulz.

18 In his text from 1866, *Generelle Morphologie der Organismen*, Haeckel defined Ökologie as follows: "Unter Oecologie verstehen wir die gesammte Wissenschaft von den Beziehungen des Organismus zur umgebenden Aussenwelt, wohin wir im weiteren Sinne alle 'Existenz-Bedingungen' rechnen können. Diese sind theils organischer, theils anorganischer Natur; sowohl diese als jene sind, wie wir vorher gezeigt haben, von der grössten Bedeutung für die Form der Organismen, weil sie dieselbe zwingen, sich ihnen anzupassen" ("By ecology,

scientist, Eugenius Warming.[19] Even so, the word describes Francé's attempt to understand the natural world as a set of relationships composed of many elements from the microscopic to the very slow. The invisibility of these elements required many different forms of vision afforded by the powerful magnifying lens or by the ability of the time-lapse camera to speed up the imperceptible movements of a plant. Even though Ernst Haeckel favoured the artist's pen over the camera for reasons that echoed Goethe's method of observation, technology was what opened up for him the possibility of seeing a dynamic and interrelated nature. Seeing the whole of nature contrasts with the tradition of specimen collecting still commonly practiced in botany. Francé places his view of the natural world in stark contrast to the botanists who persist in atomizing life into a classification system according to the principles of Linnaeus. The view of nature produced by the taxonomists resembles more a crypt than living nature in Francé's adamant words:

> Wherever he went the laughing brook died, the glory of the flowers withered, the grace and joy of our meadows was transformed into withered corpses, which this 'true botanist' collected into the folios of his herbarium, and whose crushed and discoloured bodies he described in a thousand, minute Latin terms. This was called scientific botany, and the more mummies such a register of the dead could bury in his museum the greater botanist he was held to be. (13)

Couched in the opposition between taxonomy and Francé's form of ecology is another opposition between materialism and an ensouled natural world. Elsewhere, Francé decries the limited view of Materialism, traces of which can be seen in his critique of the objectified nature quoted above. In contrast to the analytic view of nature and Materialism, Francé advocates an objective *experience* of nature in the original sense of the word. His method of getting to know plants involves venturing into nature, the forests and meadows, listening to those who interact with this world, and observing plants in their usual dynamic environments. Such an approach considers plants and others in the

---

we mean the entire system of connections of the organism to the surrounding environment, to the point, where we can count in the broadest sense all conditions of existence. These are partly organic and partly inorganic nature, as much those are of the greatest meaning for the form of the organism as we have shown before, because they are the same that force them to adapt," 286).

19 Eugenius Warming's *Plantesamd* (*Lehrbuch der oekologischen Pflanzenbeographie*, 1895) was one of the most influential books on plant geography. His concept of ecology as plant communities proposed a dynamic relation between plant and environment.

natural world as an interactive system of sentient beings with which humans can communicate in the broadest sense of the word.

Maeterlinck's and Francé's writings on plants belong to a series of discourses around nature from the early 20th century that have been neglected and denied relative to discourses around technology that fit neatly within the paradigm of Modernity as "against nature." In recent years, a field of research has been growing in an effort to address this neglect that is best represented through the 2011 anthology, *Biocentrism and Modernism*. Edited by Oliver Botar and Isabella Wünsche, the anthology identifies a series of discourses, which they call "biocentrism." While diverse and distinct, all of these discourses share a set of beliefs and themes characterized by aspects of Neo-Romanticism and Neo-Vitalism. The beliefs and themes include an "intuitive, idealistic, holistic, or even metaphysical attitude towards the idea of 'nature' and the experience of the unity of all life" (Botar 2). Francé's nature writings play a formative role in the shaping of these discourses as do Maeterlinck's nature essays. Although the goal of my research has not been to identify the novel, short stories and films as biocentric (but rather to uncover the reactions to the transformations in ideas and technology), they certainly share many affinities with Botar and Wünsche's brief definition of biocentrism: "Nature Romanticism updated by the Biologism of the mid- to late nineteenth century" (2). Since biocentrism shapes their anthology as a historical concept rather than a coherent school or a movement, the concept also helps to explain the diversity of films and short stories chosen for this book that range from canonical to relatively obscure and in styles from fantasy and horror to expressionism and new objectivity.

At the beginning of the twentieth century, there were two dominating perspectives on plant-life that informed a spectrum of reactions to the motif of the "dynamic plant". The first perspective viewed plants as living, dynamic, sentient beings, while the second opposing perspective viewed plants as immobile, mechanical, passive beings. From the latter point of view, any evidence of a dynamic plant-life serves as a metaphor for a threat to the self and to identity. In this book, I consider these two perspectives in two successive sections, focussing on film and literature. The first three chapters examine the affirmative and even utopian reception of plant life around 1900 through three examples, Lasswitz's novel, *Sternentau: Die Pflanze vom Neptunesmond*, Scheerbart's short story "Flora Mohr" (1909) and the *Kulturfilm, Das Blumenwunder* (1926). The last two chapters examine a second set of three examples that demonstrate an anxiety-filled response to encounters with dynamic plants. Meyrink's short story, "Die Pflanzen des Doktor Cinderella" (1905) is the central example of the third chapter. The expressionist films, *Nosferatu* (1922) and *Alraune* (1928), round off the last chapter.

The first chapter begins the book with the literary example of dynamic, sentient plants from Kurd Lasswitz's novel, *Sternentau: Die Pflanze vom Neptunsmond* (1909), in which an alien species of plant lands on Earth and initiates a direct human-plant communication through humanoid offspring. I argue that the function of alien Idonen in the novel is a model of interspecies communication based on the contemporaneous science, technology and aesthetics at the *fin de siècle*. As an imaginary medium, the Idonen are the utopian ideal for the potential of literature to combine objective and subjective methods for gaining knowledge about nature. Captured best in Lasswitz's coined phrase, the "scientific fairy tale," science and the fantastic complement and limit each other to form a basis for knowing nature from both physiological and psychological aspects. To create a method for subjective knowledge, Lasswitz draws on Gustav Fechner's and the Romantic nature philosophers' pan-psychism, and the literary traditions of Gothic fiction and romanticism. The physiological basis of plant-life and the human mind is derived from Lasswitz's background in science. As a result, his novel represents a paradigm shift in the early 20th century because of its critical revival of Romantic and Vitalist concepts of nature.

The second chapter continues the thread of chapter one with another literary example of dynamic plants from Paul Scheerbart's short story "Flora Mohr: eine Glasblumen-Novelle" (1909), in which the artist William Weller displays his marvellous artificial gardens of moving glass plants and lights. I argue that Weller's artificial glass plants have a twofold function in the short story as an illustration of Paul Scheerbart's aesthetics of the fantastic and as a representation of nature's dynamic force. The intent of William Weller's aesthetic program is to create original glass plants that seem alive, and yet, do not adhere to the established aesthetic conventions that connect lifelike art with realism. In this way, Weller's aesthetics are positioned as a reaction to the aesthetic demands of naturalism and impressionism, popular during the late nineteenth century. Yet, hidden within Weller's originality lies a form of imitation. In his attempt to give the flowers life through light, the material glass, and a mechanical apparatus that moves the flowers, the protagonist recreates the dynamic force of nature and illustrates a recurring argument for plant sentience—their ability to move and react to their environment. The emphasis on the dynamic quality of the glass flowers and also their inner glow, visible in the glass medium, points to a perspective of nature built upon process rather than taxonomic classification. Scheerbart's short story takes part in a revival of romanticism and provides an alternative perspective of plants from that of 19th century botanical sciences. Scheerbart's fantastical and dynamic glass plants form a part of his greater vision for revitalizing living spaces and consequently the human spirit, which he founds on the belief of a world permeated with spirit. My interpretation of his model of plant life relies on a close reading of this short story, combined with Scheerbart's many published articles as well as the theoretical texts written by the Romantics and other such proponents of the alternate botany.

The third chapter continues the thread of the first chapter, moving to representations of plants in film, which present movement as evidence of sentience. Focusing on the 1926 *Kulturfilm*, (*Das Blumenwunder*, 1926), which compared plant movement shown in time-lapse to the movements of modern dancers, the chapter interprets the mimetic interplay between the dancers and the plants as a moment of learning and communication. The film collapses the portrayal of plants as dynamic living entities from many early nature films into a feature length celebration of movement. In so doing, it reflects on the relatively new capacity of film to accelerate movement in nature through time-lapse photography as a way of making visible the dynamic force of plants. This visualization of movement through the medium of accelerated film is framed as proof of life in plants and as a means of gaining access to plant subjectivity. In addition to this evidence of plant life, the film attempts to become an opportunity for learning about the plant subjectivity by seeing the plants move and by developing a form of embodied communication with plants through dance. Simplifying the movement of plants into coded gestures opens up a means for understanding and communicating a plant subjectivity. *Das Blumenwunder* demonstrates a "plant language" through movement and then calls upon its viewers to interact with plants by imitating their movement.

The subsequent chapter on Gustav Meyrink's short story "Die Pflanzen des Doktor Cinderella" (1905) marks a shift in focus from short stories and films that view plants as benign life forms to texts in which transgressions of the assumed hierarchy of being are portrayed as monstrous. While the association of plant, human and the demonic recurs in many of Meyrink's horror and ghost stories, "Die Pflanzen des Doktor Cinderella"—an enigmatic tale of an Egyptologist who breeds carnivorous plants from human and animal organs as a scientist in a somnambulist state—is particularly fascinating for the combination of the figurative and the literal hybrid in the representation of a monstrous project of human, animal and plant interbreeding. Dr. Cinderella's monstrous, carnivorous plants, I argue, criticize both occultist practices and the materialism of the anatomical sciences for their shared result that splits the mind and the body. The reintroduction of these ancient secrets in the form of an Egyptian statue fragments the consciousness of Doctor Cinderella as well as his body, resulting in the breeding of plants that embody the principles of materialism. The plants, composed out of body parts from animal and humans, reduce life to mere vegetative growth, critiquing the analogy of machines for life. Meyrink draws on the literary tradition of the Romantic Gothic horror story, the cultural imagination of carnivorous plants, the myth of the vampire and the science of artificial life to turn his sceptical eye to both the occult and the natural sciences.

Like chapter four, the fifth chapter addresses the appearance of hybrid formations between plant, animal and human as that which undermines systems of rational knowledge and models of human identity premised on the separation from nature. In F. W. Murnau's Dracula-adaptation *Nosferatu* (1921) and Henrik Galeen's 1928 film *Alraune*—an adaptation of Hans Heinz Ewers' gothic tale of a femme fatale born of a mandrake root—the vampire and Alraune stand as obvious metaphors for the 'foreigner' and the 'New Woman,' two figures often represented as threats to the social order in Weimar culture. But the figures' more radically subversive potential resides in their association with violent plants and with the dissolution between the human and its others. This association not only disrupts the social hierarchies between genders and nationalities, but also extends this disruption to the position of humans to their fellow species. The manner with which the two films dissolve the strict taxonomical differentiation between different forms of being is seen as a perversion of the classification system but also as a defiance of the hierarchy of being that allocates to plants only the most basic of existences, which is Aristotle's continuing influence on Western culture. In both *Alraune* and *Nosferatu*, it is the lasting influence of Aristotle's definitions of plant soul (nutritive) and animal soul (perceptive) which associates the characters with the monstrous and the soulless, in addition to disrupting the hierarchy of being that places plants at the bottom. The ending of *Nosferatu* remains conservative; he is in the end destroyed, restoring the previous order of being and maintaining the taxonomy. The ending of *Alraune* retains a certain ambivalence; she is redeemed through her marriage, demonstrating mobility between species, but also a level of conservatism. The interpretation of the two films will rely on a close reading, using Aristotle's concept of the hierarchy of being and the concept of taxonomy to understand the role of the two characters, Alraune and Nosferatu.

As the filmic and literary examples will show, the many representations of plant subjectivity lie along a spectrum between two extremes. At one end, there is a move for an inclusive, dynamic identity and, at the other end, a rejection of the other and a paranoid insistence on maintaining differentiation. The inclusive end of the spectrum presents a moment for learning, communication and creativity, while the other end is laden with anxiety about human identity. In the early 20th century, new media, technology, and scientific innovations prove to be catalysts for both ends of the spectrum. They initiate a renewal in a long debate on plant sentience, intelligence, taxonomy, and dynamism by making plant movement visible.

CHAPTER 1

Flying Plants: Imaginary Media as a Model for Representing the Plant Soul in Kurd Lasswitz's *Sternentau: Die Pflanze vom Neptunsmond* (1909)

> It also takes time for the quick pace of human activity; and even God takes time to create. Time is the plant's greatest treasure, and her advantage above all other living beings is patience.
>
> *Sternentau*, 169[20]

•••

> I only think in my way [...]. But on your head—don't touch it—there sits an invisible Idone. He comes from a plant and understands what I say, and while he thinks with me, he also works with your mind and the thoughts are translated into the sounds of your language. (254)[21]

•••

> 37. The spirit appears always only in a strange, airy form.
>
> NOVALIS BLÜTHENSTAUB, NP[22]

Imagine for the moment that plants are subjective, living beings comparable to humans yet unable to communicate with us, simply because they are missing the physical capacity for spoken and written language. How then

20 "Auch des Menschen schnelles Handeln braucht Weile; und selbst derGott gewinnt Wirklichkeit nur in der Zeit. Zeit ist der Pflanze größter Reichtum, und ihr Vorzug vor allem Lebendigen ist die Geduld."

21 "Ich denke nur in meiner Art [...]. Aber auf deinem Haupte—fasse nicht dahin—sitzt unsichtbar ein Idone, er stammt von einer Pflanze und versteht, was ich sage, und während er es mitdenkt, arbeitet auch dein Gehirn mit, und die Gedanken setzen sich in den Laut deiner Sprache um."

22 "Der Geist erscheint immer nur in fremder, luftiger Gestalt."

 | DOI 10.1163/9789004327177_003

could we possibly communicate with plants? That is the central concern of Kurd Lasswitz's lengthy novel, *Sternentau: Die Pflanze von Neptunsmond* (*Star-Dew: The Plant from Neptune's Moon*, 1909). The novel takes as its premise the existence of a consciousness in plants, and then follows several paths to approach the problem of interspecies communication by drawing on discourses about the nature and appearance of the human consciousness and plant consciousness from a range of sources from science, Romantic literature and philosophy. The range of discourses places his novel on plant consciousness firmly within the heterogeneous biocentric discourses at the *fin de siècle*, by contributing to a revival of Romanic and Vitalist ideas. Lasswitz asks what would be the appropriate media to communicate the thoughts of plants, or even more precisely stated, what is the medium of thoughts? As the medium of communication, he creates the alien plant species, the star-dew, who falls to earth from Neptune's moon and produce spores that spawn intelligent beings halfway between humans and plants called the Idonen. The alien species becomes the medium of communication between human and plants, allowing both species to gain valuable insight to each other's inner life, a subjective knowledge of nature. As an imaginary medium, the Idonen represent the idealized model of how humans could communicate with plants and gain knowledge about nature in general that is not accessible through scientific method. The utopian moment of human-plant communication in Lasswitz's novel is balanced by methods of knowing nature that are already available at the *fin de siècle*. His novel proposes that the ideal of unmediated human-plant communication can be achieved through the combination of the subjective, speculative literature and the objective, fact-based science to create the "scientific fairy tale" ("wissenschaftliche[s] Märchen").

Kurd Lasswitz's education in the sciences and philosophy put him the unique position to write what we now call science fiction and what he called a "scientific fairy tale." From 1866 until his participation in the Franco-Prussian war of 1870, he studied philosophy, mathematics and physics under some renowned scholars that included the astronomer, Johann Gottfried Galle, the mathematicians, Ernst Kummer and Karl Weierstrass, and the philosopher, Wilhelm Dilthey. In 1873, he obtained a doctoral degree with his work on the behaviour of water droplets and went on to teach at a high school in Gotha, Germany.[23] There, Lasswitz continued to pursue his interest in philosophy and co-founded an intellectual exchange group called the "Mid-weekly Society of Gotha" (Mittwochs-Gesellschaft zu Gotha), in which he gave over

23 See Kurd Lasswitz, *Über Tropfen, welche an festen Körpern hängen und der Schwerkraft unterworfen sind* (Breslau: Jungfer, 1873).

sixty lectures on themes from the areas of philosophy, literature and science in addition to reading from his own texts. Lasswitz also contributed numerous articles to the "science" columns of literary journals and general interest magazines, publishing extensively in the *North and South* (*Nord und Süd*), a periodical based in Breslau and later in Berlin. Even though Lasswitz published for a popular audience, he did not abandon his research and continued to publish academic texts in philosophy and science: "Atomistik und Kritizismus, Ein Beitrag zur erkenntnistheoretischen Grundlegung der Physik" (Atomism and Criticism, A Contribution to the Epistemological Basis of Physics, 1878), "Die Lehre Kants von der Idealität des Raumes und der Zeit" (The Teaching of Kant on the Idealism of Space and Time, 1883) and the two volume "Geschichte der Atomistik vom Mittelalter bis Newton" (History of Atomism from the Middle Ages until Newton, 1890). Lasswitz intentionally wove his background in science and philosophy into his literary texts, believing that scientific knowledge of nature was fundamental to understanding ourselves and to creating a better future.

The speculative aspect of literature and the objective knowledge of science were seen to be complementary and underpinned his aesthetic theory, the "scientific fairy tale" (das wissenschaftliche Märchen).[24] In his essays "Über Zukunftsträume" ("On Dreams of the Future," 1899) and "Unser Recht auf Bewohner andrer Welten" ("Our Right to Inhabitants of other Worlds," 1910), Lasswitz defines the role of literature as the medium where science and the imagination can simultaneously complement and limit each other. In "Über Zukunftsträume," literature returns the subjective experience of nature "Naturgefühl" to the scientist's cool and objective understanding of its laws, and projects an image of the ideal human being ("eine Idealisierung des Menschen," 441). The scientific understanding of nature limits the fairy tale's flights of fancy, giving its speculative aspects a sense of authenticity and bringing the subjective experience back to the shared objective reality. The resulting "scientific fairy tale" would bring about the possibility of new and better worlds ("vor unseren Augen eine neue und höhere Welt entstehen zu lassen," 441).

Lasswitz's emphasis on incorporating science into literary texts is one of the reasons he is considered to be one of the unknown forerunners of the science fiction genre. Within the genre studies *Sternentau*, Lasswitz's novel on plants, has not been studied to any great extent; instead, his two volume novel *Auf zwei Planeten* (*On Two Planets*, 1897) has received the most attention

24 See William Fischer, *The Empire Strikes Out* for detailed analysis of the ways in which scientific method informs Lasswitz's approach to literature as based in the concept of a hypothesis.

by scholars. In the late 1970s and early 1980s, *Auf zwei Planeten* experienced something of a revival by science fiction scholars within German studies. Most studies focussed on his contribution to the science fiction genre and were invested in restoring his place within the genre and legitimizing the genre itself. Franz Rottensteiner's article "Ordnungsliebend im Weltraum: Kurd Lasswitz," and William Fischer's *The Empire Strikes Out* were particularly influential for establishing the place of Lasswitz's oeuvre within the genre and for examining the aesthetic theory underlying his work. Fischer's book does briefly touch on *Sternentau*, but only to exclude it from the science fiction genre because of its "emphasis on pure fantasy and the virtual lack of concretely described science" (59). Fischer's unwillingness to include *Sternentau* within the science fiction genre does not accurately reflect the grounding of the Idonen as an imaginary medium in the science of Lasswitz's day. His exclusion of *Sternentau* may also reflect the continuing bias against even the possibility of a plant subjectivity much less communication with it. It is not the purpose of this chapter to determine whether *Sternentau* belongs within the science fiction genre, but rather to explore science and its media as one of the possible avenues suggested by Lasswitz to overcome the communication impasse between plants and humans.

Lasswitz's explorations of a plant subjectivity in *Sternentau* and throughout his oeuvre can be partly attributed to the influence of the 19th century philosopher and scientist Gustav Fechner. Fechner is best known and revered for his ground breaking work on psychophysics, yet his philosophical treatises on the possibility of a plant and cosmic soul have been largely ignored or disregarded by scholars. Lasswitz was deeply yet critically invested in Fechner's ideas as is clear from his contributions to a revival in Fechner's ideas at the turn of the 20th century.[25] He wrote the introductions for the reprintings of *Nanna, oder, über das Seelenleben der Pflanzen* (*Nanna, or, the Soul-Life of Plants*, 1899) and of *Zend Avesta, oder, über die Dinge des Himmels und des Jenseits* (*Zend Avesta, or, on the Things of the Sky and Beyond*, 1901), in addition to an entire eponymous monograph devoted to Fechner's life work. In Lasswitz's introduction to *Nanna* (Fechner's monograph on a plant soul), he hails Fechner as visionary for his conviction that for every psychological event there is a corresponding material one. As Lasswitz reflects, this rooting of the immaterial subjective experience in the material opens up the way for Fechner to argue that there could be a certain degree of individual psychological life in plants (VIII). As with other scholars at the *fin de siècle*, Lasswitz was attracted to the proposition of a plant

25 A variety of other scholars contributed to a revival of Fechner's ideas at the turn of the 20th century, including Wilhelm Wundt, Friedrich Ratzel, Friedrich Paulsen, Bruno Wille, Wilhelm Bölsche, Eduard Spranger and many others.

soul, but criticized Fechner's method of argument and his flights of fancy, suggesting that his argument could have been strengthened by Immanuel Kant's epistemology (x).[26] Yet, Lasswitz remained committed to exploring Fechner's fundamental assertion that a form of consciousness exists in other living beings on Earth and throughout the cosmos. Lasswitz's critical reception of Fechner's ideas reflects a broader cultural turn at the *fin de siècle* to resurrect certain Vitalist and Romantic discourses, but not without reservations.

Lasswitz thoroughly examines Fechner's concept of a plant subjectivity in his monograph on Fechner from 1896, where he takes the time to outline Fechner's argument for a plant subjectivity and the potential objections he addresses. The first primary objection concerns the plant's anatomy and asserts that for a mind, both a brain and the accompanying nervous system are necessary. The second objection involves the association between movement and will, arguing that since the plant cannot move around as an animal does, it does not have a will driving the action nor the accompanying subjectivity. The two primary objections to a plant subjectivity shape Lasswitz's solution, a "Universal Translator," to the communication impasse between humans and plants in his novel *Sternentau* as well as his many other depictions of plants across his oeuvre. The "Universal Translator" refers to the fictional device from the television series, *Star Trek*, that can translate any language from any culture. In his chapter on imaginary media, Scholar Eric Kluittenberg identifies it as "a desire for technology to bridge the gap with the other," and notes its place as a recurring trope in the science fiction genre (61). Lasswitz's translator reflects this hopeful drive in science fiction, yet is also shaped through the two main doubts he needs to address and incorporate into his translator between humans and plants, in order to make his novel plausible and relatable.

Lasswitz's novel *Sternentau* has not been widely read and at over three hundred pages will not likely experience a large revival. In 1924, a section of the novel was reprinted as a hymn ("Gesang der Idonen: eine Hymne von Kurd Lasswitz") and set to music by Walter Gmeindl. Recently, the novel has been published alongside an earlier novella, *Schlangenmoos* (*Creeping Cedars*, 1884), which was also influenced by Gustav Fechner's pan-psychism. Like *Sternentau*,

---

26 The Viennese naturalist Raoul Francé also voiced similar criticisms of Gustav Fechner's method, while expressing some appreciation for Fechner's concept of a plant soul: "His book on the soul in plants is worth reading yet also confusing; the struggle of a good intuition and a good fantasy." (Sein Buch über das Seelenleben der Pflanzen ist lesenwertes und doch wieder verwirrend; der Kampf einer geniale Einsicht und einer geniale Phantasie., Das Sinnesleben der Pflanzen, 70).

the discovery of a rare plant is one of the foundations of the plot in the earlier novella, just as many elements from fairy tales are combined with didactic moments from the botanical sciences. In *Schlangenmoos*, however, Lasswitz only hints at the possibility of a plant consciousness through the "elves." In *Sternentau*, a plant consciousness is taken beyond the boundaries of fantasy and earnestly explored as a real possibility.

Sternentau's plot is structured around multiple perspectives that include the main female protagonist, Harda, the plants indigenous to Earth, and the alien plants from Neptune's moon. Each successive chapter is written from the various perspectives. The main protagonist, Harda, has discovered a new plant that she calls the "star-dew" ("Sternentau") because of the dew-like drops at the centre of what appears to be a blue flower. After meeting Doctor Eynitz in the forest, Harda learns that her star-dew is a completely unknown species that reproduces through spores rather than seeds. They undertake the study of the unknown plant species together and make the surprising discovery that the plant produces invisible, human-like creatures who are intelligent and call themselves the "Idonen." The Idonen can communicate their thoughts and those of the indigenous plants by sitting on a person's head. This direct communication with the mind also gives them the ability to influence behaviour by planting thoughts in the minds of humans. At first, it appears that the Idonen and humans could coexist peacefully until Doctor Eynitz kills and dissects one of them in the interest of science. Following this violent act, a study of human culture is then undertaken by the Idonen who come to the conclusion that the culture on earth is fundamentally divided between animals and plants. The Idonen consider themselves to be comparable in intelligence to humans and could coexist peacefully with humans, if there was not the human drive to master other species on the planet, leaving no room for the Idonen on Earth. The Idonen then decide to abandon their bodies and dissolve into the universe, leaving behind the star-dew plant. They bequeath the unique and alien properties of the star-dew plant to be used by Harda and her family's factory for the production of a compound with unusual flexibility and hardiness. It is on the basis of this plant that the family achieves prosperity and is able to solve numerous familial difficulties through marriage.

The existence of a plant subjectivity is taken as a given in Lasswitz's novel *Sternentau*. However, a large portion is devoted to countering the two main objections to a distinctive plant subjectivity, their perceived immobility and missing nervous system. The indigenous plants frame the differences in movement between plants and animals (humans) in terms of the pace and also in their differing relationships to the Earth. The plants with their slow pace of growth perceive animals to be defined by their quick movement and call them "the hurried" ("die Hastenden"), and "the treaders" ("die Treter," 30). The differences in movement are not taken to be a sign of will or agency but rather as a result of evolutionary split in the past. In contrast to plants, "[t]he animals, on the other hand, had to search for food, where they grew or ran around (because they also eat one another), and so they had to move over the ground or through the water or the air. Because of that, they became a wandering about species, hurried and unsettled, quick and powerful" (46). From the perspective of the plants, the differing rates of movement have been defined by biological necessity and, in turn, define two distinct relationships to the Earth, rather than evidence of a will and the lack of a will.

The difference in rates and types of movement define the differences between an animal and a plant subjectivity; the ways they can communicate with one another; and their history. The difference in methods of communication between plants and animals (humans) is shaped by their distinct relationships to the materiality of the Mother Earth. The plants identify their relationship to the Earth as a shared "enduring soul" ("Dauerseele"), while they characterize the human and animal soul as a "remainder of the grand Earth soul" ("Rest der großen Erdseele"), and call it an "individual soul" ("Einzelseele"). The animals have lost their "enduring soul," because they became mobile across the surface of the earth and live "divided and cut off from the eternal mother" ("geschieden und abgegrenzt von der ewigen Mutter," 46). Instead of human mobility being a sign of will and subjectivity as it has been understood since Aristotle, it has become a sign of the lost connection to the natural world by humans and animals alike. Even though the plants acknowledge that humans can communicate with one another, they still characterize humans as "silent creatures" ("schweigende Geschöpfe") and "fragmented spirits" ("trümmerhaften Geiste") who have to "eternally search for the interconnectivity of the earth consciousness" ("müssen sie auch ewig suchen nach dem Zusammenhange des Erdbewußtseins," 47). In contrast to the plants, who to human perception appear blind and dumb, it is the humans, for whom "everything remains a dark, strange secret" ("bleibt für sie alles ein dunkles fremdes Geheimnis," 47).

While the presence of a plant subjectivity occurs throughout his oeuvre, Lasswitz specifically addresses the association between subjectivity and mobility in another short story published posthumously. The story, "Die enflohene Blume: Eine Geschichte vom Mars" ("The Fugitive Flower: A Story from Mars") is told from the perspective of Martians and tells of a flower that escapes to return to her flower community. The Martians comment on the human association between mobility and subjectivity as an inability to recognize plant movement as a sign of self-expression: "They have not arrived at the idea of an individual consciousness with sensation. The reasons for this view apparently come from the fact that consciousness is only accepted, where animal-like expression, above all movement of flight or defence can be perceived, and that such processes in plants are not apparent" (190).[27] This view that plants cannot defend themselves was a common belief at the turn of the century and one that still is being discussed in popular media.[28] Plants are perceived as lacking a subjectivity and will precisely because they are perceived as immobile and unable to react to threats in their environment. Lasswitz seeks to counter this perspective through his fictional and non-fictional writings.

The plants in *Sternentau* have a distinct form of mobility that is partially shaped by the growth of the individual plant but largely by the flow of information with the biological processes as their medium. Just as we log onto the internet from a particular spot with a IP address, the plants have access to a flow of information that is both everlasting and immediate but is rooted in the material processes of the Earth:

> But between light and air, water and the earthly realm, the countless cells of plants irradiate, bathe, and touch one another in tireless interactions. All of the them tunnel their roots and rootlets in the same terra realm of the mother earth. From their great unity, where all of their exchange of powers flow together, the fine metamorphoses of the material flow back and are felt again by the cells and leaves, by the herb and tree, as the impulses of the shared origin. In this broad field of interacting chemical, electrical and mechanical tensions, every organic change propagates according to the laws, and every organ absorbs the offered energy according to its own kind. With that the plants become aware of their life. The soul of the planet, that in the genus of humankind speaks as in the fluttering of the courting butterfly, also grows in the plants as a joining together and gives them a language that is indeed incomprehensible for the human senses. (29)[29]

---

27 "Zu einem individuellen Bewußtsein mit Empfindung ist es bei ihnen nicht gekommen. Die Gründe zu dieser Ansicht liegen offenbar darin, daß man nur dort Bewußtsein annahm, wo tierähnliche Außerungen, also vor allem Bewegungen der Flucht oder Abwehr, wahrgenommen wurden, und daß solche Vorgänge an den Pflanzen nicht merkbar hervortraten."

28 For a discussion of plant sentience in popular media, see "Smarty Plants" *The Nature of Things* (CBC).

29 "Aber zwischen Licht und Luft, Wasser und Erdreich bestrahlten, benetzten, berührten sich die zahlosen Zellen der Pflanzen in unerschöpflichen Einwirkungen. Alle bergen sieihre Wurzeln und Würzelchen im gemensamen Bodenreich der Mutter Erde. Aus ihrer großen Einheit, wo aller Kräfteaustausch zusammenfließt, strömen die feinen Wandlungen der Stoffe zurück und werden wieder gespürt von Zellen und

The plants' account of their communication method resembles Ernst Haeckel's definition of ecology from 1860. Haeckel coined the term and defined it as "the entire knowledge of the connections of the organisms to the surrounding environment, to which we can also include in the broadest sense, all 'requirements for existence.' These are partly organic but also partly inorganic nature; as long as they are those that we have previously shown to be meaningful for the forming of the organism, because they are the same that force the organism to adapt to them."[30] (Haeckel, *Generelle Morphologie der Organismen*, 286). Like Haeckel's definition of ecology, plants view their communication as an interaction between themselves and the environment. Their account, however, emphasizes the dynamic and web-like aspects of how they exchange knowledge, using the Earth as the medium. Embedded in the medium Earth, the plants also have no need for a written and spoken language and no need for a nervous system capable of understanding and producing conceptual language.

Once Lasswitz has established that there is a plant subjectivity worth communicating with, he is faced with a second obstacle that the plant language is insensible to human perception and conversely that human language cannot be heard by plants. As is discussed in the third chapter on *Das Blumenwunder*, this obstacle can be solved using dance and film as a means to understand plant movement as gestures. In Lasswitz's novel, this is a problem recognized by both the plant characters and the human characters as a problem of translatability. In a chapter written from the perspective of the plants', it becomes clear that the plants are able to perceive that humans have developed a means to communicate with one another, but not with plants. The birch tree is responding to the vine, Ebah, who has expressed a longing to speak with the main protagonist of the novel, Harda: "My good Ebah! The humans speak among one another. There is no question. But who should translate our language in theirs? Don't worry yourself. It is a good thing that you can't say to Harda what she's missing, for she doesn't know in any case and she will not miss her eternal

---

Blättern, von Kraut und Baum als die Regungen des gemeinsamen Urpsrungs. In diesem weiten Felde von Wechselwirkung chemischer, elektrischer, mechanischer Spannungen pflanzt sich jede organische Veränderung gesetzlich fort, und jedes Organ nimmt nach seiner Eigenart die gebotenen Energien auf. Da werden die Gewächse ihres Lebens inne. Die Seele des Planeten, die im Genius der Menscheit spricht wie im Flattern des werbenden Falters, wach verbindend auch in den Pflanzen und leiht ihnen eine Sprache, die freilich für Menschensinne unverständlich bleibt."

30 "Unter Oecologie verstehen wir die gesammte Wissenschaft von den Beziehungen des Organismus zur umbegenden Aussenwelt, wohin wir im weiteren Sinne alle 'Existenz-Bedingungen' rechnen können. Diese sind theils organischer, theils anorganischer Natur; sowohl diese als jene sind, wie wir vorher gezeigt haben, von der grössten Bedeutung für die Form der Organismen, weil sie dieselbe zwingen, sich ihnen anzupassen."

soul."[31] (55). For the human characters, the possibility of a plant language, much less a subjectivity, remains purely within the realm of fantasy and speculation until the introduction of the alien Idonen. Before the main protagonist, Harda, has communicated with the plants through the Idonen, she imagines the possibility of a comparable subjectivity: "But the plants have completely grown on me. They aren't simply a thing, but they live and feel and every single one is something for itself. I am always imagining that if I really like a little plant, it must like me in return" (14–15).[32] Harda locates the possibility of a plant consciousness as belonging to the fantastical and identifies her feeling as going against the dominant perception of plants as objects. Lasswitz has to overcome this perception that sees humans and plants as radically different in order to create a translator between plant and human consciousness.

Lasswitz primarily solves this division through the hybrid, alien species, whose fantastical biology is capable of bridging humans and plants together as an example of continuity throughout the universe. The Idonen are uniquely positioned as still connected to the plant consciousness of their planet, Neptune's moon, even though they resemble Earth's animals more than its plants. Their ability to also connect to the plant consciousness on Earth is an example of Lasswitz's belief in the universality of natural laws. In his essay "Unser Recht auf Bewohner andrer Welten" ("Our Right to Inhabitants of other Worlds"), Lasswitz argues: "For there the basic materials and general forms of the exchange of energies are the same throughout the solar system. So it is thoroughly possible that also the organic world has been built in an analogous way everywhere because of the particularity of plasma."[33] Differences in life forms on other planets are presented as a result of the unique evolutionary path, which is shaped by the particularities of their planet and results in a biology that appears fundamentally different, but is really an alternate possibility to the Earth.

The unique evolutionary path led to what the Idonen call their organic culture, a dynamic interconnectedness, which forms the basis of interspecies

---

31 "Meine gute Ebah! Die Menschen sprechen untereinander, das ist keine Frage, aber wer soll unsre Sprache in die ihrige übersetzen? Sorge dich nicht, es ist gut, daß du Harda nicht sagen kannst, was ihr fehlt, denn sie weiß es jedenfalls nicht und wird ihre Dauerseele nicht vermissen."

32 "Aber die Pflanzen sind mir nun mal überhaupt aus Herz gewachsen. Die sind doch nicht einfach eine Sache, sie leben und fühlen ja, und jede einzelne ist was für sich. Ich bilde mir immer ein, wenn ich so ein Pflänzchen recht lieb habe, müßte mich's auch wieder gern haben."

33 "Denn da die Grundstoffe und die allgemeinen Formen des Energieumsatzes im ganzen Sonnensystem dieselben sind, so ist es durchaus möglich, daß auch die organische Welt auf Grund der Eigenart des Plasmas sich dort überall in analoger Weise aufgebaut hat."

communication. The humanoid Idonen are formed from the spores of the alien star-dew plant, yet unlike their birth plant, they can move freely across the surface of the planet. Even separated from the star-dew, they are able to communicate with "Urd," their equivalent of the Mother Earth. They call this combination of animal-like mobility and plant consciousness their "organic culture":

> The reason is that the animal life on earth, whose highest level is represented through the intelligence of humans, is equally separated from plant life in its deepest forms. While by us, through the continual switching between plant and animal generations, we, Idonen, also remain in unmediated connection to the enduring soul of the embodied world and so we are not directed by the weak contributions that are inherited from individual to the next individual. We continually take part in the reserved ancestral treasure of the whole planet. In one word, we have an *organic* culture. (305)[34]

The key characteristic of the Idonen is their ability to switch back and forth between generations of plants and animals, which allows them to synthesize knowledge from both states of being, and prevents the creation of a hierarchy of being, based upon the association of plants' with objects. Their organic culture gives a glimpse of what human culture could have been like had they not become estranged from nature and from that knowledge of their natural history. This alternative perspective acts not as a call for humans to revert to a previous, utopian, natural state of being like the noble savage Rousseau describes in his *Discourse on Inequality* (1754). Instead, Lasswitz envisions a reconnection to nature brought about by the intellectual and technological progress of civilization.

Lasswitz's vision of an organic human society is drawn from the botanical sciences at the *fin de siècle*, which regarded a particular class of plants, cryptogam, as problematic, of which the mushroom is a typical example. In the novel, Doctor Eynitz places the alien plant, star-dew, within this class of

34 "Es liegt dies daran, daß sich das Tierleben, dessen höchste Stufe durch die Intelligenz des Menschen repräsentiert wird, auf der Erde gleich in seinen tiefsten Formen vom Pflanzenleben getrennt hat, während bei uns, durch den steten Wechsel von pflanzlicher und animaler Generation, auch wir Idonen in unmittelbarem Zusammenhang mit der Dauerseele des Weltkörpers bleiben und so nicht angewiesen sind auf den schwachen Vererbungsbeitrag von Individuum zu Individuum. Wir haben dauernd teil am aufgespeicherten Erbschatz des ganzen Planeten. Mit einem Worte, wir haben eine organische Kultur." (305).

plants, which seemed to defy the reproductive conventions of plants. At the time Lasswitz wrote *Sternentau*, the cryptogam class was already considered to be an artificial construct, grouped together only through their method of reproduction, alternating generations rather than flowers.[35] The generations alternate between producing gametes and spores. A gamete is a cell that fuses with another one to reproduce sexually and is analogous to animal reproduction, while a spore can develop into another individual without first fusing into another cell (*Dictionary of Plant Sciences*, np). Lasswitz drew on this intergenerational change (Wechselgeneration) to suggest that cryptogamae could be the biological answer to the plant-human divide. In Doctor Eynitz's didactic explanation to Harda, he begins with a matter-of-fact explanation and then turns to an animal metaphor that explicitly draws similarities between animals and plants:

> Now amongst the plants, the crytogamae show something mostly similar. If we assume that our star-dew is also one, then out of the spores of this blue cup, another little star-dew plant won't grow, but rather some other plant; perhaps microscopically small or at the very least invisible, like for example, the green shoots of the fern, that one calls the pre-spores. Only on the pre-spores, the formations of two distinct genders would appear, the chicken and rooster of our example. It could also be that the whole development occurs directly in the capsule and the young chicken and rooster fly directly out of it.[36]

Lasswitz's explanation of the cryptogamae prepares the reader to accept as plausible the intergenerational change between the plant, star-dew, and the humanoid, Idonen, by associating one of the alternating generations with animal reproduction and grounding it in the cotemporaneous science. Harda's subsequent suggestion that the alien star-dew may have produced "elves"

35 For a definition of cryptogam, see Eduard Schmidlin, *Populäre Botanik* 544.

36 "Nun, unter den Pflanzen zeigen die Kryptogamen meist etwas Ähnliches. Nehmen wir an, unser Sternentau hielte es auch so, dann würden aus den Sporen dieser blauen Becher nicht wieder die Sternentaupflänzchen entsprießen, sondern irgend ein ganz andres Gewächs, vielleicht mikroskopisch klein, oder wenigstens unscheinbar, wie z.B. die grünen Täfelchen beim Farnkraut, die man den Vorkeim nennt. Erst an diesen Vorkeimen würden sich später Bildungen von zwei getrennten Geschlechtern zeigen, die Hühnchen und Hähnchen unseres Beispiels. Es könnte auch sein, daß die ganze Entwicklung sich schon innerhalb der Kapseln vollzöge und die Jungen Hühnchen und Hähnchen gleich fertig herausflögen. Und erst, wenn nachher die Hühner Eier legen, will sagen, wenn die betreffende zweite Generation ihrerseits Sporen hervorbringt, so wächst aus diesen durch Sprossung die grüne Sternentaupflanze heraus."

aids in personifying the Idonen when they are first encountered, and draws Lasswitz's science into the fantastical, where science meets the fairy tale.

The Idonen function not only in Lasswitz's novel as a model of how biology could bridge the divide between humans and plants, but also bridge that gap as a medium of communication. The point when the Idonen also play the role of a medium between humans and plants is when fantastical biology shifts into the role of an imaginary medium. By imaginary media, I am using the broadest possible interpretation: a fictional solution to an impossible desire to communicate between plants and humans, and more broadly, to resolve the human species' lost connection to the natural world. This desire to communicate with the other life forms fits neatly into one of the variants, the "Universal Translator," identified by Eric Kluittenberg's chapter on imaginary media in *Media Archaeology* (2011). The key function of a "Universal Translator" is to bring together two disparate groups and is exemplified in the television show *Star Trek*, where it allows humans to communicate freely with aliens. Like the "Universal Translator," the Idonen's function is to open channels of communication and to create the potential for new alliances on other planets. But in Lasswitz's novel, the plants have taken the place of aliens from other planets and the Idonen have taken the place of the "Universal Translator" as the direct link from the human mind to the plant consciousness as "cerebral-radiations" ("Zerebral-Strahlung," 302).

The Idonen are able to translate human thoughts into a plant's and vice versa through the materialization of the plant consciousness and the human mind. The Idonen are invisible, electric creatures who have unmediated access to both human and plant thoughts and are able to create a channel of communication through touch, sitting on a human head. The Idonen materialize the plant consciousness through translating their consciousness into the medium of the human mind, nerves and electricity. When Eynitz looks at a specimen of the star-dew's spores through a microscope, he notes that the Idonen are at a cellular level composed of the same substance as the animal nervous system:

> The new formations show forms that have until now never been observed in plants. The drawings will say nothing to you. The thing is that the whole character of the new organism has changed, also chemically, so far as I can determine. If one looks for a corresponding formation in nature, it can only be found in the animal kingdom—in the cells of the nervous substance, in the brains of humans (122)[37]

37 "Die neuen Bildungen zeigten Formen, wie sie bei Pflanzen bisher beobachtet worden sind—die Zeichnungen werden Ihnen ja nichts sagen—die Sache ist die, daß sich der

By framing the Idonen as beings composed of a "nervous substance" like animals, Lasswitz is addressing one of the key objections to a plant intelligence and one of the points used to distinguish plants from humans. In Lasswitz's monograph on Fechner, he argues that the nervous system is not necessarily required for there to be a plant consciousness, since the same processes handled by the nervous system in animals are handled by other systems in plants (53). However, a nervous system is fundamental in the novel for communication between plants and Idonen as a means to translate plant thoughts into human ones.

The nervous system not only forms the basis for translating plant thoughts into human thoughts that can be articulated as spoken language, but also acts as the interior mirror of the senses. The plant thoughts are "heard" in Harda's mind as corresponding to the sounds she uses to articulate her own thoughts. From these "heard thoughts," she can then translate the names that the "star-dew" and the "elves" have chosen for themselves. Speculating on the exact nature of the translation process, Eynitz theorizes first on the nature of the plant language as "minute changes in the pressure of the tissue" ("feinsten Druckänderungen in den Geweben"). These small changes are then translated into a "certain innervation of the acoustical centres" ("bestimmte innervationen des akustischen Zentrums"), which are in turn perceived as "certain noises, words" ("bestimmte Laute, Worte"). When there is no "appropriate expression" ("geläufiger Ausdruck"), the "unknown forms of the plant message" ("unbekannten Formen der Pflanzenmitteilung") are transformed into "original acoustic images" ("eigentümlichen akustischen Bilder"). For Eynitz, Harda's shift from the fairy tale vocabulary to the names, "Bio" and "Idonen," are the evidence of an "objective manifestation of the plant consciousness" ("objektive Manifestation des Pflanzenbewußtseins," 277).

The materialization of the plant consciousness has its correlations in controversial avenues being explored in plant studies at the *fin de siècle*. The Indian physicist and inventor, Sir Jagadis Bose, began researching plant electrophysiological responses with a belief in the continuity of life rather than the widespread view in the Western world that the plant and animal kingdoms are fundamentally divided. In the early 1900s, Bose applied his ingenious devices for detecting and recording millimetre waves to his experiments with electrical signalling in living and non-living matter. He discovered that plants have

---

ganze Character das neuen Organismus verändert hat, auch chemisch, soweit ich dies feststellen konnte. Wenn man eine entsprechende Bildung in der Natur sucht, so kann man sie nur im Tieerrech finden, in den Zellen der nervösen Substanz, im Gehirn des Menschen."

what he believed to be the equivalent of a nervous system that allows them to coordinate their movements and responses to the environment through electrical signals.

When Bose presented his research on electric signalling in plants in the early 1900s to the Royal Society and the Royal Institution in London, he was met with widespread scepticism and even hostility by his fellow researchers. Two scientists, Sir John Burdon-Sanderson and Auguste Waller, headed the group who pushed for Bose's submission to be rejected from the proceedings of the Royal Society's meeting. Both scientists were exponents of mechanistic, materialistic philosophy and had experimented on animals. Burdon-Sanderson objected to Bose's research on two counts. First, he objected to the use of the term "response" to describe the plants' movements in Bose's experiments. Second, he cast doubt on the sensitivity of the instruments Bose used to record the electric signalling of the *Mimosa pudica* (Shepherd 198). Waller's objection to Bose's work seemed to be driven more by competition than an outright rejection of his work. Waller later published his own paper on electric signalling in plants and claimed priority for his discovery of "vegetable electricity" (198). Bose's work met with such resistance from the scientific community because he challenged some fundamental assumptions on the distinction between humans and plants. In his radical conclusion, he asserted that electric signalling in plants is foreshadowed by electric response in non-living matter and that living and non-living matter respond along a spectrum.

Bose's rejection from the Royal Society's proceedings illustrates the scientific climate in Europe at the turn of the century as conservative and shaped by a belief in the hierarchy of being that places animals above plants. Lasswitz's novel reflects the cotemporaneous climate in science, through criticizing the authorities in science. Doctor Eynitz cautiously prepares their research to be presented to the authorities, gathering photographic evidence, slides, and specimens, with the belief that the authorities will otherwise see their assertion of a plant consciousness as a fairy tale: "We will hear what the authorities have to say to that. I understand that it is embarrassing for you to announce our hypothesis on the intelligence of the star-dew elf, because it would be understood as a complete fantasy" (240).[38] Harda and Eynitz believe that their research on the star-dew and the Idonen might meet with resistance from within the scientific community, because it challenges the commonly held assumption that the difference between plants and animals rests on the presence or absence of a nervous system. Where Bose attempted to demonstrate

38 "Wir werden ja hören, was die Autoritäten dazu sagen. Ich verstehe, daß es für Sie etwas Peinliches hat, unsere Hypothese von der intelligenz der Sternentau-Elfen, die hier bei uns sicher als etwas ganz Phantastisches aufgenommen werden würde, auszusprechen."

to the scientific community through experimentation, Lasswitz imagined a solution in the form of the electric Idonen, yet they both needed to contend with a materialistic philosophy that held animals and, even more so plants, to be automatons. Julien La Mettrie's *L'Homme-Plante* (*The Man-Plant*, 1748) is one such example: building on his better known and influential *L'Homme-Machine* (The Man-Machine, 1748), he compares the mechanical systems of a plant to a man's, reducing both in the process to mere material, to objects. Bose's challenge to the materialist position in plant studies continues to reverberate into the 21st century under the contentious banner of plant neurobiology.[39]

Sir Jagadis Bose's research into the electric signalling of plants and Lasswitz's electric, human-like Idonen belong to a broader network of discourses about nerves and electricity during the "age of nervousness."[40] By the mid-19th century, nerves were understood to be material means of communication between the external world and the mind. First described by Wilhelm Erb in 1860, the disease of neurasthenia or "tired nerves" had become by the 20th century a mass phenomenon and had crossed social classes from the bourgeoisie to the working class. At first, neurasthenia was believed to be treatable through electrotherapy that would "regenerate" the nervous system (Killen, 65). By the 1890s, the effectiveness of such treatment was being called into question, attributing the "cure" to the suggestibility of the patients (76). In his lecture *Über die wachsende Nervosität unserer Zeit* (*On the Growing Nervousness of our Time*) from 1893, Erb described the illness as a particularly modern phenomenon, a reaction to the overburdening of modern life (7). In Lasswitz's novel, the electric Idonen not only address this "overburdening" of modern life, but they also reflect the perception of nerves as the means to communicate directly with and influence the mind.

To translate plant consciousness into human thought, the Idonen have to induce a nervous illness in the human subject, Harda. Her nervous illness occurs when she communicates with the plants through the Idonen and shares many affinities with the symbolic status of neurasthenia from around 1900. In the introduction to his book, *The Cult of the Will* (2008), Michael Cowan argues that nervous illness experienced a paradigm change around 1900,

39 The name, plant neurobiology, is itself seen as contentious, let alone the contentions of the field. The field refers to the study of plant behaviour in the physiological processes by which a plant can sense and communicate with its environment. The use of the term "behaviour" to describe this interaction is responsible for the controversy.

40 For a more extensive discussion of the network of discourses about nerves, see Michael Cowan, *The Cult of the Will* (University Park: Pennsylvania State University Press, 2008) and Andreas Killen, *Berlin Electropolis: Shock, Nerves, and German Modernity* (Berkeley: University of California Press, 2006).

when it came to be associated with the cult of the will. He writes that nervousness was associated with "psychic passivity" in the bourgeois imagination; that is, "a pathological inversion of the normative function of autonomous subjectivity" (8). Harda reluctantly experiences a similar state to the "psychic passivity" of a nervous patient, when she communicates with the plants through the Idonen for the second time:

> Suddenly, she felt on her forehead the cool breeze that she recognized from earlier. She became frightened and reached with her hands to her head, while letting the book fall from her hands. But her hands were driven immediately backwards as if they had been shaken by an electrical shock. Should she let herself indulge in this seizure? She felt the power of her self-determination dwindle and waited once again to see and hear a plant before her. (213).[41]

Harda's passivity as she awaits the plant communication resembles the nervous patient, but curiously reframes the nervous illness as a transformation from human to plant subjectivity, the main difference lying in the emphasis on the collective versus the individual. Contact with the Idonen causes Harda to relinquish control of her body and its mobility to the Idonen in order to gain access to the web of knowledge within the collective plant consciousness. Just as when a person has lost any outward appearance of thought or intention, they are called a vegetable, Harda's body, in the moment she relinquishes control to the Idonen, also begins to resemble a plant.

While the materialization and translation of plant consciousness derives largely from discourses about nerves, the Idonen are also contextualized within a familiar genre for readers, the ghost story. Lasswitz's novel shares with early ghost stories, the Gothic fiction (*Schauerroman*) from the turn of the 18th century, the framing of ghosts as a means to communicate between the material body and the immaterial spirit, their appearance as invisible and insubstantial, and their association with electricity and suggestibility. In his article on Friedrich Schiller's unfinished novel *Der Geisterseher* (*The Ghost Seer*), Stefan Andriopoulos argues that Schiller's novel was intended as an "essay" (Aufsatz) to "warn against the dangers of superstition and enthusiasm

41 "Plötzlich empfand sie an ihrer Stirn den kühlen Hauch, den sie von früher her kannte—sie erschrak und griff mit den Händen nach ihrem Kopfe, indem sie das Buch fallen ließ. Aber ihre Hände fuhren sogleich zurück wie von einem elektrischen Schlage durchzuckt. Sollte sie den Anfall über sich ergehen lassen? Sie fühlte die Kraft ihrer Selbstbestimmung schwinden und erwartete, wieder eine Pflanze vor sich zu sehen und zu hören."

while investigating the possibility of spiritual communications according to unknown laws" (66). According to Andriopoulos, the apparitions stand in for the invisible reach, power and deceptions of secret societies (73). Lasswitz's novel *Sternentau* shares with Schiller's novel its warnings against pure speculation but inverts the threat of the invisible, secret society, the Idonen, to a mostly benevolent species. Schiller's influence on Lasswitz's novel also includes his speculations on the medium of the spirit. Lasswitz appropriates Schiller's concept of a "mediating force" (Mittelkraft) from his *Philosophy of Physiology* (1779) as "partly spiritual" ("teils geistig") and "partly material" ("teils materiell," 77) that "lives in the nerves" ("wohnt in den Nerven," 80).

It is significant that the first sighting of the Idonen occurs within a chapter called "Ghosts" ("Gespenster"), immediately triggering an association with Gothic fiction in the reader. This contextualizing of the Idonen situates them as a "mediating force," as part body and part spirit. The watchman helps the reader to interpret the role of the Idonen in the narrative by labelling them ghosts: "No, no, little miss. I am completely sober, but I have really seen it. There were lights between the trees; there are souls, who fly about! [...] They are much bigger [than fireflies] and a great deal less bright, only a little, dull shimmering; one could not clearly recognize it." (84).[42] Composed of light, the ghostly Idonen recall the visual codification of ghosts as light, flying beings, and halfway between materiality and immateriality. The watchman's subsequent description links the Idonen to the plants and their material counterpart: "[...] in the middle of the meadow, there stood a thorn bush and there peered out of it something light—" ([...] mitten auf der Wiese, da stand ein Distelstrauch, und da guckte etwas Helles heraus—," 85). The Idonen are thus framed as the spirit of plants yet mobile like the human spirit.

Lasswitz appropriates the coding of ghosts from Gothic fiction to signal their place as a medium of communication between different states of being. In Gothic fiction, the primary function of ghosts was often to convey a message from the afterlife to the living or to warn of some imminent danger. The second, non-staged apparition in Schiller's *Geisterseher* predicts an event—the exact time of the cousin's death—across time and space. Like Schiller's apparitions, the Idonen are a medium to communicate with the beyond and are able to flout the rules of time and space. As invisible beings and with a direct connection to plant and human thoughts through their "cerebral radiation"

42 "Nein, nein, Fräulein. Ich bin ganz nüchtern, aber ich hab's wirklich gesehen. Es sind Lichter zwischen den Bäumen, das sind Seelen, die dort 'rumfliegen! [...] Sie viel größer [als Glühwürmchen] und viel weniger hell, nur so ganz matt schimmernd, man konnt's nicht deutlich erkennen."

("Zerebral-Strahlungen"), the Idonen can communicate over vast distances and connect different states of being. It is no coincidence that they are connecting the mobile, human consciousness with a consciousness that resonates with features common to the world of the dead. Like the dead, the plants often occupy a similar symbolic space as objects or bodies without a soul. The emphasis on disparate states of being dominates Harda's first communication with the plants:

> And the elves float onwards and lower themselves onto the head of the girl. They stopped in the soft, elastic silk of the hair. They are talking with one another in a language, which is not spoken by people. Yet, what they are saying quivers afterwards in the mind of the slumbering girl and forms human thoughts there. She believes she hears the voices of the forest and unknown worlds open secretive gateways in front of her brightly shining eyes. (109–10)[43]

There are at least three different states of being described in Harda's first contact with the Idonen, which all exemplify in-between-ness. The first is the ghost-like state of the Idonen, who exist in between states and can communicate between the "living" Humans and the "dead" plants. The second, Harda asleep and dreaming, creates a state of receptivity in Harda. Her body, no longer under control of her conscious mind, resembles the immobile plants, and in such a state can access the network of plant communication. The last state, the plant consciousness, resembles a network of communication, however, one that is hidden underneath an exterior that appears dead, the plant.

Her communication with plants while in a dream-like state confirms German Romanticism as yet another essential source for Lasswitz, and one that justifies his choice to use literature as the medium for combining the subjective experience with the objective one. The first reference to German Romanticism was not, however, Harda's communication with plants, but the star-dew, a blue flower with dew drops at the centre. The blue flower has become symbolic for German Romanticism and signals here Lasswitz's

43 "Und die Elfen schweben weiter und lassen sich auf das Haupt des Mädchens nieder, dort ruhen sie in der weichen, elastischen Seide des Haares. Sie reden zu einander in einer Sprache, die von Menschen nirgends gesprochen wird, und doch, was sie sagen, bebt nach im Gehirn der Schlummernden und wirkt darin Gedanken nach Menschenart. Stimmen des Waldes glaubt sie zu vernehmen, und unbekannten Welten öffnen geheimnisvolle Pforten vor großleuchtenden Augen."

appropriation of the project of Universal Poetics (Universalpoesie) through his combination of science with literature as the "scientific fairy tale." The blue flower also flags Lasswitz's novel as reflection on the nature of consciousness and in a highly self-reflexive manner, a reflection on that reflection. While it is not my intention to explore in great length the Romantic aspects of Lasswitz's novel, his reliance on a particularly Romantic understanding of literature explains his choice of the novel as a medium to explore a plant consciousness. As Martha Helfer argues, the romantic writer and philosopher Novalis saw literature as "capable of representing the unrepresentable, the pure ego" (238). Harda's first communication with the Idonen can be seen as an encounter with the "pure ego." Her encounter echoes in many ways the opening sequence of Novalis' novel *Heinrich von Ofterdingen* (1802), when Heinrich has a vision of his inner self. In a dream, Ofterdingen sees a blue flower that turns into the face of a young women. Yet before she is able to speak, he awakens from the dream and begins his quest for self knowledge. Yet, unlike Ofterdingen, Harda is able to hear the voice of the flower and the forest through the Idonen. Instead of representing the unattainable, the blue flower embodies an internal paradox in Lasswitz's novel, language as the medium of human subjective experience and capable of fully giving voice to the non-human. Yet, within Lasswitz's choice of literature is contained the particularly modernist realization of its failure to capture the totality of the human experience and to represent the non-conceptual plant consciousness through conceptual language.

Out of the three literary works discussed in this book, Lasswitz's is the one that grapples most with the constraints of a language-based communication and conversely its potential. This contradiction manifests in the novel as an imaginary media, the Idonen, who can translate between conceptual language and the language of plants, blooming. Lasswitz was not alone in his recognition of the constraints of language. Many writers at the turn of the century also grappled with the limits of language to fully express the subjective experience. Prominent symbolist writer and naturalist, Maurice Maeterlinck, described the inability of language to express the depths of experience in his essay *The Treasure of the Humble* (1896), using a metaphor that compares language to a drop of water in an ocean of experience (61). This concern with language was not limited to the symbolists but captures a widespread tendency at the turn of the century that represents an increasing scepticism of language to represent embodied experiences. Writer Franz Kafka also noted the limits of language to depict experiences outside of conceptual language in his short story, "Ein Bericht für eine Akademie" ("A Report for an Academy," 1917), in which a monkey describes his life before he learned to speak as a blank space.

This blank space is encountered by Harda when she attempts to communicate with the plant Ebah without the Idonen's help. She senses the forest as only a child would, who, believing that plants are living beings, senses self-expression in the "sunny shimmery colours" ("sonnigem Farbenschimmer"), the "dark wave of the shadow" ("dunklem Schattenwink"), the "soft rustling of leaves" ("leisen Blätterrauschen") and the "pure, still forest's breath" ("reinen, stillen Waldesatem," 150).

Even when problematically depicting the conversations of the plants through a medium inaccessible to plants, language and literature, the novel self-consciously reflects on the ability of the writer to represent plants through conversations between Eynitz and Harda's uncle, Geo Solves. Lasswitz consciously places his novel within a tradition of texts, the fairy tale, that personifies plants and gives them a voice to question whether literature could achieve more than that: "does the writer only lend his mouth to the plants as the fairy tales once did [...]" ("der Dichter nur der Pflanze seinen Mund leiht, wie das Märchen von je getan hat[...]"). Or is it possible to fundamentally alter the relationship of humans to the natural world: "or whether his work forces us to hear the plant itself as a child of the mother earth corresponding to our contemporary knowledge about nature ("oder ob sein Werk uns zwingt, die Pflanze selbst zu hören als Kind der Mutter Erde, entsprechend unserm gegenwärtigen Wissen über die Natur," 360–61). Lasswitz invites the reader to reflect on the success of his attempt to make the voice of the plants convincing and whether the reader now views plants as fellow beings. The point of contextualizing the novel within a literary history, from Gothic fiction, Romanticism and fairy tales, which each represent spirits, souls and consciousness in various ways, is to explore literature as a medium to express the nature of consciousness.

The conversations in *Sternentau*, also theorize the role of literature as a medium for bringing together subjective and objective experiences. The Idonen characterize the problem as a Cartesian split in the being between fantasy (the subject) and the external forms (the object). The individual subjective experience is defined as the "eternally animated, free formation" ("ewig rege, freie Gestaltung"), while reality is defined as the surrounding environment ("das was sie umgibt," 307). Through the conversation between Geo Solves and Eynitz, we learn that the role of literature is to reunite fantasy and the external world as the "scientific fairy tale."

In Solves and Eynitz's conversation, literature is positioned as the medium to explore the personal, subjective experience, because of its ability to externalize the psychological experience, consisting of emotions, impressions, and dreams. Instead of following an agenda, Solves believes that the writer gives form to his or her inner life: "The writer can only find a form for that what moves his inner life, his soul, in order to make him feel alive in the beautiful appearance of it, which is truth and life for all who take part in it." (360).[44] This subjective aspect is what gives the medium the power to explore ideas that are intuitive (such as the existence of plant-human communication) rather than objectively provable: "But apparently the experience of the connection between plants and humans can be so powerful in the writer that he tries to portray it. Obviously, not to teach it, but rather to give a feeling of a lasting endurance that requires a form." (360).[45] In Lasswitz's introduction to Fechner's *Nanna*, he clarifies the danger of knowledge only gained through the subjective experience: while the writer gives voice to the "pounding waves of infinity in the human heart" (Wogenschlage des Unendlichen im Menschenherzen, 3), there is the danger of mistaking the subjective experience for a universal truth: "But it remains hidden to them, how much of what they believe to be the breaking waves of the ocean is actually only the beat of their own heart or the calming rhythm of their own personal certainty of faith."[46]

The capacity for literature to give form to the fleeting subjective experience is best captured through Ebah's blooming at the end of the novel. In Michael Marder's *Plant-Thinking*, he traces the association between blooming and the spirit to the German Idealists (61). In Lasswitz's novel blooming as the highest strata of spirit is one of its many functions, but the one that receives the final word. Before Ebah's blooms, Harda frames his act of blooming as a function of literature to give form to the spirit by quoting a poem that expresses her feeling of oneness with the universe: "You lead a line of living beings / going by in front of me and teach me / to recognize / in the quiet bush, in the air and water / my brothers" (386).[47] Although Harda can no longer "hear" the Ebah through the Idonen, both the objective reality of his blooming is described and then his subjective experience of blooming is given form. Ebah, "whispers in *his language* [...] Shadows, I am blooming and the wasp is flying! Like a holy blessing, it grows in me. I am with you. I am with all of you. I am in the forest!

44 "Der Dichter kann nur eine Form finden für das, was seine Seele im Innern bewegt, um es lebendig zu machen in dem schönen Schein, der Wahrheit ist und Leben für alle, die daran teilnehmen."

45 "Aber wohl kann in einem Dichter das Erlebnis jenes Zusammenhangs von Pflanze und Mensch so mächtig werden, daß er es darzustellen versucht; selbstverständlich nicht um zu lehren, nein, sondern um einem Gefühle Dauer zu verleihen, das nach Form verlangt."

46 "Aber ihnen bleibt es verborgen, wieviel von dem, was sie für die Brandung des Weltoceans halten, nur der Pulsschlag des eigenen Herzens, nur der beruhigend Rhythmus der persönlichen Glaubensgewissheit ist."

47 "Du führst die Reihe der Lebendigen / Vor mir vorbei und lehrst mich meine / Brüder / Im stillen Busch, in Luft und Wasser / kennen."

But completely alone, I am once again for me, for me myself." (386).[48] The "translation" of Ebah's blooming into a written text reflects literature as a privileged media to express the subjective spirit in nature. At the *fin de siècle*, literature was also one of the few time-based mediums, capable of tracing the striving of a plant toward the light and its eventual blooming.

Ebah's final pathos-filled act of blooming is not an unqualified expression of Lasswitz's imagination but has been balanced by alternative narratives from aesthetics and evolution. The cultural context of blooming has been theorized throughout the novel in the conversations between the plants as both an aesthetic expression of unity and a biological necessity. It is, therefore, unsurprising that Lasswitz chose the horse chestnut tree as the advocate for the aesthetic function of flowers. Known for its beautiful flowers, the horse chestnut tree can lay claim to its distinguished place within human culture.[49] In response to the spruce tree's indifference to flowers, the horse chestnut tree raises the point of aesthetic value as an expression of the individual and a point of intersection between humans and plants:

> That is really nonsense, [...] We have elaborated that all so prettily with blooming and that is the most beautiful and elegant part of us. That is the aesthetic culture, as the people say, that is the refinement that binds us with animals and humans. If we were to get rid of that, we would sink further and further into the masses. But that also doesn't work. (48)[50]

Blooming becomes a metaphor for the progress of the plants that can only occur from the recombination of individual plants. It is also a form of self-expression that can be recognized by humans as an act of individuation, and represents an intersecting point between human and plant cultures.

48 "flüstert in seiner Sprache: [...] Schattende, ich blühe und die Wespe fliegt! Wie ein seliges Heil wächst es in mir. Ich bin bei dir, ich bin mit euch allen, ich bin im Walde! Aber ich ganz allein bin noch einmal für mich, für mich selbst."

49 In many of the books on the language of flowers, the horse chestnut was associated with luxury. See J. J. Grandville, *Les Fleurs Animeé* 56. and Carey, Lea and Blancard, *The Language of Flowers* 13.

In a garden journal from 1873, the horse chestnut is referred to as belonging to a group of trees with the most beautiful flowers (*Neue allgemeine deutsche Garten- und Blumenzeitung*, 65).

50 "Das ist ja Unsinn,[...] Das haben wir alles so fein ausgearbeitet mit dem Blühen, und das ist das Schönste und Vornehmste an uns. Das ist die ästhetische Kultur, wie die Menschen sagen, das ist Verfeinerung, das verbindet uns Tier und Mensch. Wenn man das abschaffte, würde man immer mehr in die Masse versinken. Aber es geht auch gar nicht."

The horse chestnut's description of the aesthetic communication is qualified in turn by the moss's description of blooming as a biological necessity. If there were no blooming, there would be "no progress" ("keinen Fortschritt"); the descendants would be "too similar, uniform and after a certain pattern" ("zu gleichartig, eintönig, schlabonenhaft," 49). With the addition of the evolutionary explanation, the example of blooming demonstrates how the concept of progress can be reflected in a narrative form that bases subjective experience in objective science. In the last scene of Ebah's blooming, literature has taken the place of the imaginary medium, the Idonen, as the space for this interspecies communication.

The interplay of objective and subjective knowledge of nature is exemplified through effects of names on perception and vice versa. The path from a purely imaginative concept of nature to a combination of subjective and objective can be seen in the many variations of names for the Idonen. Harda first conceives of the mysterious plant and its humanoid offspring through fairy tales and calls them the "star-dew" and "elves." Eynitz perceives the plants and the Idonen through the lens of science and calls them their Latin names, "cryptogam" and "gametyphte." The final set of names are self-determined, "Bio" and "Idonen," and reflect the plants' and their offspring's sense of self. The three distinct set of names correspond to the subjective perspectives of Harda and the Idonen, and to the objective perspective of the scientist, Eynitz. The shift to an integrated knowledge of the natural world can be seen when Harda shifts from using vocabulary from fairy tales to using the self-determined names.

The effect of names on perception is then illustrated through the disenchantment of the glowing moss. When Doctor Eynitz and Harda encounter the glowing moss, his first reaction is emotional and sympathetic, calling the moss "a fairy tale" (Ein Märchen), but his second reaction is to label the plant with its official Latin name, "*Schizostega osmundacea*" (20). Harda provides the interpretative clues for Eynitz's reactions to the glowing moss, identifying the first as emotive and connective: "So one feels, don't you think? And that should not be sympathized with? Absolutely nothing should be noticed, that it doesn't also strive with us to the higher power, the light? ("So fühlt man's, nicht wahr? Und das soll nicht mitfühlen? Sollte gar nichts merken, daß es mitstrebt wie wir nach dem großen Gotte, dem Lichte?", 20). After his second reaction, she provides the didactic explanation: "Yes, glowing moss [...] I know it. Light bending cells light it with their own leafy green. But you have deflated the magic." ("Ja, Leuchtmoos [...] Ich weiß es. Lichtbrechende Zellen beleuchten sich ihr eignes Blattgrün. Aber Sie haben den Zauber gelöst", 22). With the mastery of scientific language also comes the limiting of emotion and fantasy. The idea that science contributes to the disenchantment of the world

(Entzauberung der Welt) is not an original one, insofar as it refers the increasing rationalization of society, and in which the people, animals and plants are increasingly objectified and controlled. The sociologist, Max Weber, predicted it as an eventual result of the practice of science as the magic of the primitive world view would be gradually explained away (Weber 537). However, the way the novel is more forgiving than Weber and integrates fantasy as fundamental to knowledge of the natural world, viewing magic as simply a primitive form of science.

Harda and Eynitz's botanical studies are evidence of what the Idonen have called the "technical" culture and contrast with their "organic" culture. Recall that the Idonen's organic culture refers to the interconnected relationship of the intelligent species to their planet. The differences between "organic" and "technical" are defined by how each obtains knowledge about the natural world on their planets. While the Idonen maintain their link to their planet and its reservoir of the natural knowledge through their psychic connection to plant life, humans need to use other means to overcome the communication impasse and gain knowledge of their "Mother Earth." These other means are what the Idonen call the "technical culture" ("technische Kultur") and are defined by the "path of intelligence" ("Weg der Intelligenz," 306). The path of intelligence is further defined by "work" ("Arbeit"), "knowledge" ("Erkenntnis"), "effort" ("Mühe") and "adversity" ("Not," 306–7). As the word "technical" also implies, the tools and methods also serve the path to the intelligence, and aid to a great extent Harda and Eynitz's examination of the Idonen. A few examples have already been referenced and include microscopes, photography, chemicals and prisms. In a hierarchy reminiscent of Kant, the technical culture is privileged by the Idonen over the "formation of fantastical myths" ("wunderlichen Mythenbildungen," 306) and "emotions" ("Gefühle," 307). The fantastical myths are perceived to be capable of creating only a close facsimile of the Idonen's unified experience by circling around it ("herumdeuten"). Emotions also prove to be an unreliable method of gaining knowledge, leading to destruction ("vernichten") rather than "unity" ("Einheit"). The same privileging of reason over emotions is prevalent throughout Lasswitz's writings. In his essay, "Über Zukunftsträume," he reflects on "technical culture" ("technische Kultur") as the way to better and increase the knowledge of nature ("Naturerkenntnis"), to improve the living standards of humans ("die Lebensbedingungen verbessert" and to bring peace and understanding to international relations ("die Nationen einander schätzen lernen," 435).

The idealistic aspect of Lasswitz's belief in the possibility of technology and other media can be seen in the role of microscopes, photography and the other optics in *Sternentau*. His choice of technology and media fit with their contemporary use of them, and are framed in the novel as an expansion of the capacity of the senses to perceive natural processes in the service of objective knowledge. The microscope is used to distinguish whether the Idonen cells are similar to an animal's or a plant's as we saw from an example discussed earlier. At the time that Lasswitz had written *Sternentau*, the microscopic world was just gaining widespread attention. The scientist, Ernst Haeckel, had drawn attention to the mysterious forms it reveals in his *Kunstformen der Natur* (*Art Forms in Nature*, 1899). As is discussed in chapter three of this book, microscopic images of blood cells are used in the film *Das Blumenwunder* (1926) as a means for visualizing commonalities between humans and plants. Lasswitz's readers would have also had access to books on the application of microscopes that were intended for the broader public.[51] In Lasswitz's novel, microscopes are one of the many media that thematize how innovations in technology can make previously unseen worlds visible. The utopian potential of microscopes, photography and optics is to also prove what is not yet provable, the spirit of nature.

The idea that technology can reveal what is hidden to our senses is particularly apparent in the role of the opera glass with Nicol prisms in the place of lenses in the novel. A Nicol prism was invented in 1828 by William Nicol in Edinburgh for enhancing clarity of microscopes by eliminating certain waves of light. In the novel, the opera glasses have a similar purpose to the microscope and play a similar role to the x-ray glasses from later science fiction: they make visible what is invisible: "If one looked through a glass, that with the exception of the leafy green rays, all other colours were nearly but not fully dimmed [...] it was possible to recognize the Idonen web as a dark red drawing." (Wenn man durch ein Glas sah, das mit Ausnahme der Blatt grünstrahlen alle andern Farben stark, aber nicht völlig, abblendete, [...] das Idonengespinst als eine dunkelrötliche Zeichnung zu erkennen, 336).

There was already a rich tradition in photography for visualizing spirit before Lasswitz's novel. Called "spirit photography," William Mumler created his first photograph of a ghost around 1870. The double exposed portraits of a living person and a "spirit" built on a visual code for ghosts as hovering between visible and invisible, material and immaterial. The spirits in Mumler's photographs appear washed out and less substantial than the image of a living person. In Lasswitz's novel, photography echoes the function of Mumler's spirit photography to make the spirit-like Idonen visible: "For Harda came to the idea, to now make a photographic shot in the bright light and furthermore by dimming the red to green rays. It could be that the invisible 'elves' in this way would be photographically imprinted. Perhaps, the plate would show

51 See Friedrich Merkel, *Das Mikropskop und seine Anwendung* (München: R. Oldenbourg, 1875).

more than the eye could see" (213).[52] Beyond visualizing the spirit, the photograph as a document of the natural world could provide objective knowledge of the Idonen outside of the anecdotal "sightings."

The role of the media and technology in Lasswitz's novel extends beyond their ability to enhance the senses and make the invisible visible. They also serve to enhance the individual's claim to authority by making the subjective experience an objective reality. In Lasswitz's article "Naturnothwendigkeit und ihre Grenzen" ("Absolute Necessity and her Borders"), he defines knowledge ("Erkenntnis") as the ability to make a subjective perception of the natural world a general reality (59). In his novel, the photographs and the microscopic preparations become the medium of this general, objective reality and replace the immediate experience of the Idonen. Without the evidence, their "hypothesis" ("Hypothese") of the Idonen's intelligence would be certainly believed to be a "total fantasy" ("ganz Phantastisches") by the authorities within the botanical sciences (240). When the Idonen destroy all telling evidence of their existence, the role of the media as proof of intelligence in plants becomes painfully obvious, leaving Harda and Eynitz no other choice than to abandon their decision to publish their results (357). With no evidence, their knowledge of plant subjectivity remains an unsubstantiated subjective experience of the world. Harda's uncle Geo Solves expresses this early failure as a promise of science and media:

> And from there, it will be successful sooner or later to spread awareness in our time that there is not only a plant soul, but also that messages can be sent from the plant soul to the human soul regarding natural entities. And the main thing is however always that all earthly beings learn to understand one another in the shared grand consciousness of the divine. (360).[53]

Alongside the optimism in the future of science and media, the limits of science and specifically botany to fully know the other are being qualified.

52 "Da kam Harda auf den Gedanken, jetzt bei dieser hellen Beleuchtung eine photographische Aufnahme zu machen und zwar mit Abblendung der Strahlen von Rot bis Grün. Vielleicht zeigte die Platte mehr, als das Auge sehen konnte."

53 "Und da wird es früher oder später einmal gelingen, unsrer Zeit zum Bewußtsein zu bringen, nicht bloß, daß es eine Pflanzenseele gibt, sondern daß es selbst bei einer Mitteilung der Pflanzenseele an die Menschenseele mit natürlichen Dingen zugehen kann. Und die Hauptsache ist doch immer, daß wir Erdenwesen alle uns verstehen lernen im gemeinsamen großen Bewußtsein des Göttlichen."

The foremost limit of the sciences is precisely the same one that is their greatest strength, their ability to analyze the natural world as objects. While it gives scientists the ability to rely on the knowledge acquired as a universal law, it also does not give them access to subjective experience of other living beings. In his introduction to his book on Gustav Fechner, Lasswitz defines the limits of science as that which can be measured and weighed, and can determine the laws of nature and the relations of living beings. The scientist is limited to the "certain path of experience" ("sicheren Bahn der Erfahrung") and carefully avoids going into the "world of dreams," into the "innermost factory, out of which the pulse of the world takes its way out, in the deep of the life of the soul" ("innerste Werkstatt, von welcher der herzschlag der Welt seinen Ausgang nimmt, in die Tiefe des Seelenlebens," 2). The separation of emotions from the practice of science results in the natural world being treated as composed of objects rather than living beings.

The tendency within the natural sciences and specifically botany to view plants as objects turns into an overt criticism of this practice through Doctor Eynitz's treatment of the plants. In the chapter called, "The Botanist" (Der Botaniker), Harda encounters Doctor Eynitz in the forest with a small collection of plant specimens laid out on a table in front of him: "On the table, there was a straw hat, a few plants released with their roots, a little knife, scissors and two small, glass bottles. The owner of these utensils was so busy that he didn't notice Harda's arrival. With a deeply bowed head, he observed a small leaf that he held with tweezers through a magnifying glass." (13).[54] The killing of a few plants for the purpose of study only begins to become problematic when seen from the perspective of the plants. Provoked by Eynitz's collection of plant specimens from the forest, Ebah, the vine questions the inequity he perceives in the human-plant relationship. Ebah asks his fellow plants and trees, "Down there, the herbs are saying that the Treaders have cut off and dug out many of them. He also took a few of the foreign plants, my quiet fosterling. We saw her lying on the table. Should we tolerate that?" (31).[55] The vine's indignation at Eynitz's complete disregard for the plants' lives points to the

54 "Auf dem Tische befanden sich ein Strohhut, einige mit ihren Wurzeln ausgelöste Pflanzen, Messerchen, Schere und zwei Glasfläschchen. Der Inhaber dieser Utensilien aber war so eifrig beschäftigt, daß er Hardas Kommen nicht einmal bemerkt hatte. Er betrachtete mit tief herabgebeugtem Kopfe durch die Lupe aufmerksam ein mit der Pinzette gehaltenes Blättchen" (13).

55 "Unten erzählen die Kräuter, der Treter habe viele von ihnen abgeschnitten und ausgegraben. Auch von der fremden Pflanze, meinem stummen Schützling, nahm er einige. Wir sahen sie ja auf dem Tische liegen. Sollen wir das dulden?" (31).

fundamental obstacle within the sciences to furthering the knowledge of the natural world. As the plants later comment, humans call everything that is not human "nature," and they "believe it is dead, and soulless" ("Sie nennen's Natur, aber sie halten es für tot, für unbeseelt," 33).

The criticism levelled at the practice of botany echoes a call for the reassessment of botanical sciences at the turn of the century by the naturalist and writer Raoul Francé. Eynitz exhibits the classical behaviour of a botanist that Francé criticized so intently in *Das Sinnesleben der Pflanzen* (*Germs of Life in Plants*, 1907). Francé specifically criticized the Linnaean tradition of taxonomy that encouraged the perception of plants as both dead and soulless:

> This obsessed spirit of a man was possessed by a true mania for registration who even divided up his friends into categories and subdivisions. Through his formidable authority, he preserved from out of the night of scholasticism the dried up spectre of what he called the *verus botanicus*, the true Botanist. Wherever he went, there died the laughing meadow, there wilted the glory of flower, the adornment and joy of our fields changed into dried bodies, which the *verus botanicus* piled up in the folders of his herbarium and whose mis-coloured, pressed bodies, he then described in thousands of quibbling, latin diagnoses. That is called scientific botany. (8)[56]

The specific criticism levelled at Linnaeus argues that the study of plants as dead specimens reduces them first to individual objects and, second, is an inadequate method to understand the plants as living, interactive beings. In his examination of the plant specimens, Doctor Eynitz is only able to perceive the plants through what is already known and labels them accordingly as cryptogamae, but he cannot determine whether they are sensing beings.

While the plants in Lasswitz's novel are in the end willing to tolerate this treatment (whether by choice or not), the Idonen are not. Their resistance to

---

56 "Dieser [Linneaus] von einer wahren Registratormanie besessene Geist, der sogar seine Freunde in Kategorien und Subdivision einteilte, erhielt durch seine ungeheure Autorität bis in unsere Jugendein aus der Nacht des Scholastizismus heraufgetrochenes Schemen am Leben, das er den verus botanicus, den wahren Botaniker nannte. Wohin er trat, da erstarb dielachende Aue, da verwelkte die Blumenpracht; die Zier und Freude unserer Fluren verwandelte sich in getrocknete Leichen, die der verus botanicus in den Folianten seiner Herbarien aufhäufte, und deren mißfarbenen, zerdrückten Körper er dann in tausenden spitzfindiger, lateinischer Diagnosen beschrieb. Das hieß wissenschaftliche Botanik" (8).

being treated as objects of study reveals an underlying critique of the practice of dissection in science. When the Idonen find out that Eynitz has captured one of the Idonen and killed it with chloroform, they react with outrage. Their subsequent heated discussion interrupts Harda's telepathic conversation with the vine Ebah and she overhears their immediate responses: "'Imprisoned! Killed, killed before he could give himself a name.' 'Who? Who? Who did it?' The voices melted into one another. 'The person, who here—' 'In some house, he holds Stefu in prison—' 'There, the person sits, who was so often with him—' 'What do you want from Harda?' 'They must be annihilated!'" (256).[57] The Idonen eventually decide to free the remaining captured Idonen and destroy all visual evidence of their existence on Earth. While too late for the dead and dissected Idonen, their reaction causes Harda to see their practice of science from the perspective of the Idonen: "It was natural that they had to regard the humans as their violent enemies, as criminals" (257).[58] Yet both she and Eynitz state several times that they could not have acted any differently (257, 281). Lasswitz's critique of science for its objectification and cruelty towards other living beings stops just short of advocating for better treatment of animals. Instead, the world view of science as objective and cool is presented as inevitably leading to the objectification of others. As a result, the practice of science is intrinsically limited from addressing the subjective aspect of living beings. In a conversation between Eynitz and Harda, these limits of scientific study become an ethical question and reveal the hierarchy of being as necessary for its practice:

> Are these elves [Idonen] really intelligent beings? How can we be allowed to then treat them as simply objects of experimentation? They are positioned over the animals. They are comparable to us—Say we preemptively assume that—It would be possible—Am I then allowed to simply kill them in order to study them? It goes against my beliefs to use force against them—yet however, what should we do? (257)[59]

---

57 "'Gefangen! Getötet, getötet ehe er sich einen Namen geben konnte—' 'Wer? Wer? Wer tat es?' Die Stimmen gingen durcheinander. 'Der Mensch, der hier—' 'In jenem Hause hält er Stefu gefangen—' 'Da sitzt ja der Mensch, der so oft bei ihm war-' 'Was wollt ihr von Harda?' 'Sie müssen vernichtet werden!'"

58 "Es war natürlich, daß sie die Menschen für ihre tätlichen Feinde halten mußten, für Verbrecher."

59 "Sind diese Elfen wirklich intelligente Wesen, wie dürfen wir sie dann einfach als Objekte des Versuchs behandeln? Sie stehen über denTieren, sie gleichen uns—nehmen wir das einmal vorläufig an, möglich wäre es—darf ich sie dann schlechtweg töten, um sie zu

The answer comes from Lasswitz's masterpiece novel, *Auf zwei Planeten* (*On Two Planets*, 1897), in the form of the Martians' advanced technology. As with the Idonen in *Sternentau*, the Martian species have a relationship with plants characterized by benevolence. The Martian technology has developed to the point that they no longer rely on plants for nutrition and have removed themselves from being ethically implicated in the death and suffering of plants:

> Stones in Bread! Protein and Carbohydrates from rocks and earth, from air and water without the means of the plant's cells! That was the art and science used by the Martians to emancipate themselves from the lowly cultural position of agriculture and to place themselves as the unmediated sons of the sun. The plants served as aesthetic pleasure and as protection for the moisture in the earth, but their suffering was not allowed. (254)[60]

The concern for the plants' suffering is correlated with a belief in a plant subjectivity that can suffer and that technology provides the solution to this suffering. In *Auf zwei Planeten* as with his other novels and short stories, the alien species stands for an idealized humanity, where science and technology are not tools of mastery, but where they aid in improving the living standard and ethical position of humans in relation to other species and knowledge of the natural world. The potential of technology to alleviate suffering is crucial to Lasswitz's later novel *Sternentau*, by articulating the goals of technological progress and emphasizing the current treatment of plants and animals at the hands of humans as potentially cruel, a question that is being explored by scholars such as Monika Bakke.[61]

A second answer comes in the form of the capacity of literature to imagine the subjective state of the other. Lasswitz believed that in order to experience empathy, it is critical that we recognize ourselves in the other. In his essay

---

studieren? Mir widerstrebt es, dann mit Gewalt gegen sie vorzugehen—und doch, was sollen wir tun?"

60 "Steine in Brot! Eiweißstoffe und Kohlenhydrate aus Fels und Boden, aus Luft und Wasser ohne Vermittlung der Pflanzenzelle!—Das war die Kunst und Wissenschaft gewesen, wodurch die Martier sich von dem niedrigen Kulturstandpunkt des Ackerbaues emanzipiert und sich zu unmittelbaren Söhnen der Sonne gemacht hatten. Die Pflanze diente dem ästhetischen Genuss und dem Schutz der Feuchtigkeit im Erdreich, aber man war nicht auf ihre Erträge angewiesen. Zahllose Kräfte wurden frei für geistige Arbeit und ethische Kultur, das stolze Bewusstsein der Numenheit hob die Martier über die Natur und machte sie zu Herren des Sonnensystems."

61 See Monika Bakke's article "Art for Plant's Sake? Questioning Human Imperialism in the Age of Biotech." in *Parallax*. (18.4 (2012)): 9–25.

"Unser Recht auf Bewohner andrer Planeten," he writes that, "the reader can only be captivated in the points where his own interests and experiences are being included."[62] He goes on to explicitly state: "Because of that, the poetic has to continually anthropomorphise. Otherwise, their personalities and characters would be incomprehensible to us" (172).[63] It is the same perspective that creates the opportunity for Eynitz and Harda to acknowledge the problematic aspects of their treatment of the Idonen. The effect of personification is one of the central themes of writer Alfred Döblin's novella, "Die Ermordung einer Butterblume" ("The Murder of a Buttercup," 1913). After the main character lops off the head of a buttercup, he struggles with an immense amount of guilt stemming from his personification of the flower. While the personification of plants and animals contains several problematic aspects such as eliding difference, it also highlights the differences in moral standards used to measure plants and animals. While Lasswitz does not use his novel *Sternentau* to call for complete rehaul of the way science is practiced, he does use the literary device of personification to reflect critically on science as a method to know nature.

In addition to literature as a hybrid medium of subjective and objective experience, the female body as a medium and femininity as an aspect of it are also proposed. Lasswitz's novel takes a unique and progressive position on the role of women that both subscribes to stereotypical beliefs on the role of women and inverts them. Harda as a woman is placed as closer to plants and to nature in general than Eynitz, which gives her an advantage over Eynitz. This close proximity to nature is represented in the novel through blending her body in with the forest: "The full ash blonde hair over her forehead glinted in the broken light of the forest in a green shimmer" ("Das volle aschblonde Haar über der Stirne glänzte in dem gebrochenen Lichte des Waldes in grünlichem Schimmer," 9). Her proximity to nature is repeatedly reinforced as in this next example: "On the white dress, reddish rays of the setting sun were playing, hair and shoulders shimmered green in contrast under the reflection of the broad canopy of the beech, and the brown eyes glowed impatiently from out of the shadow of the hand on her face [...]" (27).[64] When the conflation of nature

62 "Denn der Leser kann nur dort gefesselt werden, wo er an seinen eignen Interessen und Erlebnissen gepackt wird."

63 "Die Poesie muß daher stets anthropomorphisieren, sonst würden ihre Persönlichkeiten und Charaktere uns unverständlich sein."

64 "Auf dem weißen Kleide spielten rötliche Strahlen der niedergehenden Sonne, grünlich schimmerten dagegen Haar und Schultern under dem Widerschein des breiten Buchenlaubes, und die braunen Augen leuchteten ungeduldig aus dem mit der Hand beschatteten Gesicht..."

and the female body is extended to Eynitz's perception of her, it becomes associated with emotion: "[…] and in the reflection of the beech tree's canopy, the full, wavy hair itself shimmered in golden green, all life and soul at once."[65] The conflation of the plants and Harda's body reveals that it is on the site of the female body that human beings can come closer to the plants. Harda's body anticipates the role of Flora, the protector of flowers, as a mediator as well as the embodied communication of dance and film discussed in chapter three through the film, *Das Blumenwunder*.

There is another dimension to the physical connection of Harda to the plants that opens the opportunity for plants to become more human-like. Through her body Harda's body becomes an example of a living flower rather than simply a personified flower: "There she stood like a flower that has come alive." ("Da stand sie wie eine lebendig gewordene Blume." 28). In this reference to a history of comparing flowers and women, there are a collection of associated vegetal traits bound up with the reference that include passivity and an inferior status, but Lasswitz inverts those attributes to position Harda as open to the plant subjectivity. By animating a flower, Harda also embodies the main idea of the novel that plants are living beings, children of mother earth as are humans. If Harda's role were limited to this traditional conflation of femininity and nature, then there would be nothing different about Lasswitz's concept of femininity and the vegetal kingdom. Yet, Harda's relationship to the natural world in many ways confounds the traditional views of femininity as both capable of an intellectual and an emotional engagement with the natural world.

Harda's relationship to nature is characterized by her love of the natural world, but also by her desire to gain knowledge of it through the methods of scientific inquiry. It could be argued that Harda is merely following a tradition of acceptable science for women. Yet, she also proves herself to be equal to the task of understanding difficult concepts. 19th century botanical works written for female readership were often adjusted to accommodate the perceived lesser abilities of women to reason.[66] In a conversation with Eynitz, when discussing some of his more technical experimental findings, she needs to reassure him that she understands: "You needn't excuse yourself, Mr. Doctor; I am prepared for academic lectures. […] Didn't I tell you that I possess the

65 "[…]und im Gegenschein des Buchenlaubes das volle, wellige Haar selbst wie in goldigem Grün schimmerte, alles Leben und Seele zugleich."

66 For a clear outline of the relationship of women to botany, see "Botanizing Women" *Clandestine Marriage*.

completion certificate of a high school?"[67] She also justifies her desire to study botany as a sincere wish to learn more: "I really meant it. Father could do without me until just now. But please, go ahead. You wanted to say in any case that a separation of the genders arise, masculine and feminine spores" (121).[68] Her use of technical language to describe the process of reproduction reveals her perception of them as objects of scientific study. Her desire to study botany is also set in conflict with her traditional duties of maintaining peace and order in the household, clear from the reference to her father's need of her. She frames her duties within the household as constricting, from which she longs to be free. Set in contrast to the household duties are her moments studying nature or being within nature. Encapsulated in Harda is a complex relationship between a subjective experience of nature and an objective understanding of it, but also the conflicts facing many women from the early 20th century.

Lasswitz's literary texts have met with less attention than they deserve from scholars and the public alike. The repression of his texts during World War II by the National Socialists and his ponderous prose that gives his work a dated feel are a few of the reasons. Yet, his works still have much to offer 21st century readers who are facing an environmental crisis of epic proportions. In the Idonen as an imaginary medium and a model of integrated communication, he shows the way not only in the literal sense of human-plant communication but also in the metaphorical as a way to urge the human species as a whole towards improving the way we can know the others amongst us who seem radically different. Lasswitz supports his utopian vision of unmediated communication in contemporaneous discourses of science, technology and humanities to form a realistic and practical method to achieve his vision. Furthermore, his reliance on these discourses is coupled with his knowledge of how these discourses are in many ways a revival of past ideas coupled with the most recent discoveries. This places his text within a heterogeneous set of discourses at the turn of the century that has been referred to as "biocentrism" and is defined as a critical revival of Romantic and Vitalist concepts. Lasswitz's novel *Sternentau* has much to offer the 21st century by envisioning a better, alternative relationship and laying out a method for creating that relationship, grounded in science, made possible through technology, and given flight through words.

---

67 "Sie brauchen sich nicht zu entschuldigen, Herr Doktor, ich bin auf akademische Vorträge vorbereitet. [...] Habe ich Ihnen nicht mitgeteilt, daß ich das Reifezeugnis eines Realgymnasiums besitze?"

68 "Es ist ganz ernstlich gemeint. Vater konnte mich nur bis jetzt nicht entbehren. Aber bitte, fahren Sie fort. Sie wollen jedenfalls sagen, daß eine Trennung der Geschlechter auftritt, männliche und weibliche Sporen."

CHAPTER 2

Animating Glass: Representing the Elusive Plant Soul in Paul Scheerbart's "Flora Mohr: eine Glasblumen-Novelle" (1909)

There, everything is moving. There, nothing is at all dead.[69]

WILLIAM WELLER

The narrator of Hugo Hofmannsthal's short prose sketch, *Die Rose und der Schreibtisch* rescues a dying rose from the snow and lovingly places it in a vase on his desk for one last chance to breathe.[70] An artificial, porcelain rose, in possession of a well-established place on his desk, mocks the narrator for his bad taste in placing the living rose next to the artificial. Hofmannsthal's sketch seems to uncover a contradiction from the point of view of literary history represented by the beauty of the living rose and the preference for the artificial as expressed by the flower, an "old viennese inkwell" (alt-wiener Tintenzeuges). As represented in the conflict between the narrator's care for the living rose and artificial rose's scorn, the *fin de siècle*'s infatuation with vital plants seems to run counter to the prevailing currents of Decadence and Symbolism, which tended to valorize a cult of extreme refinement and artificiality that eschewed the vitality of living nature. This can be seen in both the heightened artificiality of the poetics as it is privileged over an organic and looser form, and in the themes of disease and decay. Yet, if the two roses in Hofmannsthal's story can be read as a glimpse into prevalent themes in decadent literature, then part of its pathos is the way in which decadent literature reveled in images of plants existing on the verge of death. Indeed, the passing from the natural to the artificial was a prevalent theme of the garden in decadent literature, as analyzed by literary critics such as Carl Schorske and seen in such examples as Stefan

69 "Da bewegt sich Alles. Da ist gar nichts tot" (515).

70 See Hugo von Hofmannsthal, "Die Rose und Der Schreibtisch," in *Gesammelte Werke in 10 Einzelbänden. Erzählungen, erfundene Gespräche und Briefe, Reisen* (Frankfurt am Main: Fischer, 1979), 443.

 | DOI 10.1163/9789004327177_004

George's dying park from his famous poem "Komm in den totgesagten Park und Schau" ("Come in the Park Declared Dead and Look," 1897). George's park is a realm utterly opposed to the dynamism of Fechner's plant soul, one whose last vestiges of "green life" (grünes Leben) were to be transformed by the reader into an "autumn image" (herbstliches Gesicht).[71]

Little wonder, then, if the worlds depicted in turn-of-the-century literature and painting were so often populated by artificial flowers. And yet, such artificial worlds were not always as devoid of vitality or as removed from nature as one might at first assume. One writer who attempted to blend the cult of artifice with an interest in vital nature through images of plants was the science fiction author Paul Scheerbart. Although Scheerbart's short story "Flora Mohr" (1909) has received little attention in scholarship on the author, the text is highly relevant to the present study because it offers a good example of the kinds of utopian representations of plant life and plant souls that I discussed in my introduction, which emphasize the benefit from seeing plants as living beings. The story of one inventor's creation of fabulous artificial gardens constructed of glass, light and moveable machinery, "Flora Mohr" might appear at first glance as a literary heir to the kinds of artificial decadent worlds made famous in novels such as Joris-Karl Huysmans' *A rebours* (*Against Nature*, 1884). But as I will show in what follows, even as Scheerbart explicitly opposes his protagonist's mechanical flowers to any naturalistic imitation of nature, he nonetheless insists that his dynamic flowers possess both "life" and "soul," and he even draws on imagery reminiscent of Goethe to describe their movements. I argue that the fantastical flower spectacle depicted in Scheerbart *should*, in fact, be seen as an effort to channel and imitate nature. However, "nature" appears here not as a collection of static and classifiable species familiar from botanical treatises, but rather as a *living, creative* force—precisely the kind of vital force that cotemporaneous research into plant life had revealed. Scheerbart's imaginary spectacle of mechanical plants represents an attempt to use technology in order to cultivate and liberate such this creative force, and in this sense, the short story also depicts an effort to reconcile technology and nature. Moreover, as I show towards the end of the chapter, this endeavor also helps us to situate Scheerbart's imaginary plant spectacle within a broader current of turn-of-the-century visual culture—stretching from Loïe Fuller's light and electricity dances to early experimental cinema—that sought to convey the dynamism of nature rather than representing static objects. If plants show up as the referents for so many of these dynamic spectacles, this is surely

71 See Stefan George, *Sämtliche Werke in 18 Bänden*. Band 4: Das Jahr der Seele (Stuttgart: Klett-Cotta, 1982), 12.

no accident; for by the time Scheerbart wrote his short story, the plant had been rediscovered as an "object" teeming with creative force. It is this creative force that the protagonist of "Flora Mohr" attempts to recreate in his garden of moveable glass flowers.

"Flora Mohr" consists of two narrative levels moderated by a narrator called Münchhausen, who is an adaptation of the historical and fictional character.[72] Münchhausen is speaking to a Japanese audience and recounting his experience of some of the fantastical gardens created by Australian inventor William Weller. The inset narrative is Münchhausen's talk, describing what he saw on Weller's estate and the discussions he had with Flora Mohr, Weller's niece who is temporarily living with her uncle. A visiting, rich Indian nabob also contributes to the discussions on art and nature as he browses through Weller's gardens, looking for an acquisition. According to Münchhausen, Weller has constructed a palatial greenhouse that gives the impression of a "big botanical garden" with many "natural flowers" outside (492). The building serves as a workspace and as an exhibit for Weller's many animated glass plants and flowers, arranged as landscapes and gardens and illuminated with coloured lights. Münchhausen's tour of Weller's compositions begins in a rotunda with statuesque and oversized glass flowers in the centre and on the walls. From there, he enters Weller's experiment room that consists of a floor inset with many glass lenses and varying arrangements of flowers underneath. They breakfast in a room with glass fruit hanging from trees reminiscent of a garden from *1001 Arabian Nights*. The other compositions include a lake filled with growing glass flowers; a garden with moving flower beds; a jungle-labyrinth with moving mirrored-walls and grottos; a rotating tower-panorama that depicts various stations of a flower's life; and finally a kaleidoscopic comet flower. The title character Flora Mohr acts as a contrast to her artist uncle by repeatedly arguing that the natural flowers are far more beautiful than artificial ones and insisting on sincerity and practicality. Her criticism heightens the beauty of the flowers for the visiting nabob and causes him to buy the finest of Weller's creations, the "comet flowers" for a large sum of money. Münchhausen's account of Weller's art to the Japanese audience is repeatedly interrupted by his requests for food, drink and rest as well as descriptions of the ocean view, and comments from his audience and Clarissa, Münchhausen's German travelling companion. After

72 Hieronymus Carl Friedrich von Münchhausen (1720–97) was a German nobleman, who fictionalized his experiences in the Russian military, producing outrageous and extraordinary stories carried out by the fictional Baron von Münchhausen. The link between the 18th century Münchhausen and Scheerbart's is clear from a reference to the Russian campaigns at the end of "Flora Mohr".

Münchhausen finishes his account of what he has seen, both he and Clarissa leave the Japanese garden in an automobile at dusk.

Scheerbart's contribution to art, architecture and literature has been well-established through scholarship that concentrates on his place in the development of science fiction and fantasy literature (Ege, Patsch),[73] his role as a precursor to Expressionism (Raabe),[74] his roots in Jugendstil (Ruosch)[75] and his impact on architecture (Bletter, Stuart). Scholars often gravitate to Scheerbart's masterpiece novel *Lesabéndio* from 1913 as a rich text open to a myriad of interpretations and a ripe place to examine his visual aesthetics, world view, and satirical wit.[76] In architecture circles, he is still well-known for his visions of colourful, brightly lit glass architecture as laid out in his manifesto *Glasarchitektur* from 1914. However, less attention has been paid to Scheerbart's use of 19th century spectacle including the phantasmagoria, the panorama, and the glass palace as well as his reception of the increasingly popular forms of moving image spectacles. The impact of Romantic writers and philosophers of nature on Scheerbart's world view has also been well documented, by tracing back his concept of a world soul to the influence of Fechner's panpsychist theory.[77] However, in favour of his vision of the cosmic

---

73 According to Ege, Scheerbart bases his fantasy aesthetic, a blend of science fiction and fantasy, on the epistemology of his day, seeing the role of fiction and science to broaden the realm of the possible (224). Cornelius Partsch addresses the difficulty of fitting Scheerbart's work, specifically Lesabéndio, within the development of the science fiction genre.

74 Paul Raabe identifies Scheerbart, "als größter Phantast der deutschen Literatur, als genialer Außenseiter der Jahrhundertwende und als Vorläufer des Expressionismus" ("as biggest visionary of German literature, as genialist outsider of the turn of the century, and as a precursor to Expressionism," 54).

75 In one of the few references to Scheerbart's plant imagery, Ruosch identifies the arabesque and the serpentine as characteristic of both *Jugendstil* and Scheerbart's plants (73).

76 The complexity of the novel leaves it open to a myriad of interpretations including "eine Erweiterung unserer Gefühls- und Vorstellungsschranken ins Kosmische" ("an expansion of the limits of our feelings and imagination into the cosmic," (Adelt 222), a "social-anarchical culture utopia" (Bär 75)." and a, "tausendfarbiges Monumentalmosaik von berauschend-gestuftem Rhythmus" ("thousand coloured monumental mosaic of intoxicating, sliding rhythm"), a "kosmophilosophische Bekenntnisschrift" ("cosmo-philosophical confession") (Grätzer 224), and the "utopische Bild einer geistigen Gestirnwelt" ("utopian image of a spiritual mind world") from the "reinsten unzweideutigenn Erscheinung der Technik" ("purest, non-ambivalent appearance of technology") (Benjamin, Sur Scheerbart, 230).

77 See for example Müzeyyen Ege's *Das Phantastische im Spannungsfeld von Literatur und Naturwissenschaft im 20. Jahrhundert,* which I discuss in more detail below.

soul, scholars have often passed over Scheerbart's more familiar and earthy depiction of the world soul, the plant. Prevalent in "Flora Mohr," the dynamic plant also runs throughout Scheerbart's oeuvre, opening up the possibility for new readings of his work.

"Flora Mohr" was conceptualized as a sequel to Scheerbart's novel, *Münchhausen und Clarissa* (1906), and published separately in 1909 as well as in the later collection of Münchhausen short stories, *Das Große Licht*, from 1912. "Flora Mohr" stands out from his other narratives on account of the way in which it develops his aesthetics specifically through the visualization of a plant soul. In "Flora Mohr," Scheerbart chose to elaborate on the flower and plant imagery from *Münchhausen und Clarissa* as opposed to the novel's other fantastical images, including caverns deep within the earth, fantastical creatures at the bottom of the sea and life in the cosmos. As seen from Münchhausen's description of an exhibition from the World Fair in *Münchhausen und Clarissa*; glass, water, and light already form an integral part of the plant spectacles in Scheerbart's works:

> Clearly, the story of the evening was almost supernatural from all the colossal lighting effects. The lighting came first out of the water lilies. And how that appeared in the many glass flowers—you can easily picture that yourself. Naturally, the eighteen big captive balloons above were transformed into light-art-flowers. And the wires above were made into countless light-vines. And then, on the lake there was a colossal fountain composition that gloriously mirrored the above story in certain places. It finally appeared as if you could believe that all these coloured water streams were flowers. In addition, there was a brilliant balloon music just underneath the balloons. Various musicians sat just underneath the air balloons in gondolas. And in the middle, you saw the baton of the bandmaster moving up and down as a huge light-diamond-sceptre. And with that the flower world simply fell on the lake, so that finally only the gigantic balloon flowers were reflected on the lake. And then it was suddenly completely dark above. And afterwards a play of light flowers of enchanting movement began underneath the surface of the lake.[78]
>
> SCHEERBART, *Münchhausen und Clarissa*, 69

78 "Selbstverständlich war die Geschichte des Abends bei all den kolossalen Beleuchtungseffekten fast überirdisch. Die Beleuchtung kam zunächst aus den Seeblumen selber heraus. Und wie das bei den vielen Glasblumen wirkte—das können Sie sich natürlich leicht ausmalen. Natürlich wurden auch wieder die achtzehn großen Fesselballons oben in Lichtkunstblumen verwandelt. Und aus den Drähten oben wurden unzählige

The dynamism in the World Fair exhibit in Sydney is characteristic of Weller's compositions as are the use of light, glass and water to enhance the sense of fluidity and weightlessness. Münchhausen's description of Weller's interior lake from "Flora Mohr" repeats many of the same details from the image in *Münchhausen und Clarissa*:

> On the water swam twelve water lilies—completely colourful water lilies. And they suddenly began to glow colourfully, so that the dew work also became completely colourful. [...] We sat ourselves in a small row boat. [...] The water lilies dispersed colourful colour-clusters—like coloured projectors. [...] William asked me to look in the depths of the lake—and there I saw how coloured flowers grew slowly up. And the flowers grew up out of the water and glowed. They also glowed in the depths of the water.[79]
>
> SCHEERBART, *Flora Mohr*, 502

The two compositions share an emphasis on dynamism and fluidity that is also characteristic of Scheerbart's images of the cosmos, suggesting that movement is fundamental to Scheerbart's world view. To visualize the plant soul, Scheerbart is also clearly leaning on the spirit iconography from the 19th century which depicted ghosts and spirits as hovering between visibility and invisibility, material and immaterial.[80] For example, spirit photography made

---

Lichtschlinggewächse. Und dann gabs auf dem See, der die obere Geschichte an einzelnen Stellen prachtvoll spiegelte, eine kolossale Fontänenkomposition, die schließlich so aussah, daß man alle diese sprühenden bunten Wasser für Blumen halten konnte. Und dazu gabs oben unter den Ballons eine großartige Ballonmusik. Sämtliche Musiker saßen oben in den Gondeln unter den Luftballons. Und in der Mitte sah man den Taktstock des Kapellmeisters als ein großes Lichtdiamantenscepter auf- und absteigen. Und dabei zerfiel die Blumenwelt auf dem See allmählich, sodaß schließlich auf dem See nur die gigantischen Ballonblumen gespiegelt wurden. Und dann ward es oben plötzlich ganz dunkel. Und danach entstand unter der Seeoberfläche ein Funkenblumenspiel von entzückender Beweglichkeit."

79 "Auf dem Wasser schwammen zwölf Seerosen—ganz buntfarbige Seerosen. Und die begannen plötzlich bunt zu leuchten, sodaß das Tauwerk auch ganz bunt wurde. [...] Wir setzten uns in einen Kahn. [...] Die Seerosen streuten bunten Farbenbüschel aus—wie bunte Scheinwerfer wirkten die Büschel. [...] William bat mich, in die Tiefe des Sees zu blicken—und da sah ich, wie bunte Blumen langsam emporwuchsen. Und die Blumen wuchsen aus den Wassern heraus und leuchteten. Sie leuchteten auch in der Tiefe des Wassers."

80 Tom Gunning in his article on spirit photography and phantasmagoria from the 19th century discusses the way in which these forms signal spirit as that which is less substantial

visible the ghost as a transparent image, which indicates its state of being as insubstantial. While also characteristic of Jugendstil, the dynamic plant imagery stretches farther back to Romantic writers and philosophers, in particular to Gustav Fechner, whose perception of the natural world is very close to pantheism.[81] Scheerbart's preference for glass, colours, water and light reflects this intent to represent the creative force flowing within the entire physical world.

Although plant imagery is not as pervasive in Scheerbart's oeuvre as his depictions of life in the cosmos, he did repeatedly return to the dynamic plant throughout his writings. In addition to *Münchhausen und Clarissa* and "Flora Mohr," he describes multi-coloured flowers growing deep within the moon in his novel from 1902, *Die große Revolution* (*The Big Revolution*). The flowers, which the moon inhabitants smoke, anticipate the dynamism and weightlessness of Weller's glass compositions (17).[82] Another type of flower from the same novel, the lightening-flower, possesses both a flower language and a glass-like effect that are later echoed in "Flora Mohr" (45–6).[83] As with the flowers in "Flora

than living beings. See Gunning (To Scan a Ghost: the Ontology of Mediated Vision." *Grey Room*. 26.4 (2007): 99).

81 See Introduction for a discussion of Fechner's concept of a cosmic soul.

82 "Diese Blumen, die in allen Farben irisieren und opalisieren, sind hauchartig dünne Fächergebilde und kolossalen bunten Eisblumen ähnlich; aber die Blumen in den Rauchergrotten sind nicht einseitig—sie können sich nach allen Seiten entfalten—werden weite Spitzenblüten und Strahlendüten mit Schaumranken und haarfeinen Adern, die sich kräuseln—zitternd und glühend." ("These flowers, which iridesce and opalesce in all colours, are similar to breath-like thin images of fans and to colossal coloured frost, but the flowers in the smoker grottos are not one-sided—they can unfurl themselves to all sides—become broad pointed-blossoms and light rays with vines of sponges and hair-fine arteries that curl—shaken and glowing.").

83 "Die Blitzblumen nahmen unterdessen immer größere Formen an und wuchsen jetzt auch am hellen Tage in den sammetgrünen Himmel hinein. Und die Mondleute glaubten, daß diese Blumen eine große Blumensprache sprächen. Und die Weltfreunde deuteten diese Sprache natürlich zu ihren Gunsten. »Der Mond selber«, sagten sie, »will, daß wir seine Glasgefilde näher kennenlernen, denn sonst würden die Blumen nicht so glasartig wirken.« Das Glasartige und Durchsichtige, das jetzt den Blumen vielfach eigen war, schien nun allerdings die Meinung der Weltfreunde nur zu bestätigen. Oft flatterten die riesigen steifen Blätter der Blitzblumen wie kolossale irdische Libellenflügel in der Luft herum. Nur ein paar Sekunden lebten die geisterhaften Blumen aber sie ließen sich doch photographieren. In den Photographien konnte man erst die ganze Farbenpracht und die entzückende Aderzeichnung der Blattwandungen erkennen und genießen." ("The lightening flowers took on among those always bigger forms and grew now also in the light of the day into the velvet green sky. And the moon people believed that these flowers would be speaking a grand flower language. And the friends of the world naturally interpreted

Mohr," technology reveals the beauty in nature, when the sensational range of colours in the lightening-flowers are made visible by capturing their brief appearance in the lasting medium of photography.[84] As with the spirit photography from the nineteenth century, photography visualizes the ephemeral. In another novel also from 1902, *Liwûna und Kaidôh. Ein Seelenroman* (*A Soul Novel*), the two main characters fly over a forest that consists of massive flowers with an intoxicating scent (20).[85] The plant imagery from these early novels shares with those in "Flora Mohr" the dynamism, ethereal bodies, and use of colour, suggesting a certain constancy to Scheerbart's view of nature. The inclusion of plants in his depictions of cosmic life reflects the high esteem he held for flowers as well as the connection he saw between the cosmos and flowers.

Starting with *Münchhausen und Clarissa*, technology, in addition to the transparency, colour and dynamism, becomes an even more integral part of Scheerbart's later plant imagery. In the description of his attempts to create a perpetual motion machine from 1910, *Das Perpetuum mobile. Die Geschichte einer Erfindung* (*The Pertuum Mobile. The Story of an Invention*), he integrates mechanical elements and the art of projection into moving plants and intends to power the garden with the perpetual motion machine, which he has named "Perpeh":

---

this language to their convenience. 'The moon itself,' they said, 'intends for us to get to know its glass images closer, for otherwise the flowers wouldn't have the glass-like effect.' The glass-like nature and transparency, which in many ways are one with the flowers, appeared to now confirm in any case the opinion of the friends of the world. Often the huge stiff leaves of the lightening flowers, flapped around like colossal earthly dragonfly wings in the air. The ghostly flowers lived only for a few seconds, but they let themselves be photographed. It was first in the photographs that one could recognize and enjoy the range of spectacular colours and the enchanting drawings of veins on the leaf walls.").

84 Photography is paradoxically both valued and devalued as a medium in "Flora Mohr." It is valued for its ability to present a life-like representation, yet this capacity also has the effect of devaluing the original artwork. Weller does not allow his artwork to be photographed because it would prevent him from receiving the optimum price. As a result Münchhausen must rely on language, even as it inadequate to illustrate for his Japanese audience the beauty and originality of Weller's art.

85 "Berauschender Duft steigt da den Beiden in die Nase. Der Himmel ist hell und weiß wie Kreide. Doch unten blühen Riesenblumen—so hoch wie Berge—Blütenkelche so tief wie Täler—Staubfäden wie schwankende Leuchttürme. An einer langen Mauer hängen Weintrauben, die so groß sind wie dicke Bündel aufgeblasener Luftballons Ringsum ein Urwald aus Riesenblumen!" ("Intoxicating scent rose up there into both of their noses. The sky is light and white as chalk. Yet under blooming massive flowers—as high as mountains—calyxes so deep as valleys—filaments like swaying lighthouses. On a long wall hung grapes that are as big as thick bundles blown-up air balloons. Surrounded by a jungle of massive flowers").

> A garden whose parts are movable. Transportable hedges, transportable terraces. And particularly: transportable flower beds. Light in the evenings through latern slides that are being lit from underneath. Flower basket garlands hanging on chains. Huge masts with blossoming flowers in baskets of earth, that are movable to pull up and down—and can also rotate around the masts. The flowers have to hang out of the baskets. Movable flower walls from wire. Walls for protection against the wind. The movable must win the upper hand in the garden. Plants on big frames that can be moved with Perpeh. Moving Lighting. Swimming flower beds in the pond. Automatically moved big structures with glimmering glass pieces. (np)[86]

The combination of plants and technology culminates in his novel *Lesabéndio*, where phosphorescent plants serve as lights for mushrooms and other fungi that have been specially manipulated to increase the harvest and to serve as a food source. The plant imagery is informative for the picture it provides of the way in which technology, media, nature and spirituality are all intertwined in Scheerbart's aesthetic vision.

Technology and science underpin many of Weller's glass flowers compositions, opening up the possibility for him to represent pure creative energy, fundamental to Scheerbart's vision of spiritual art. Throughout "Flora Mohr," the technology behind the movement of his glass compositions is conspicuous even when it is hidden. When on the interior lake, Weller is defending his glass compositions against Flora's condemnation of the artificial flowers as dead and empty, he states: "Do you see how the saphire-blue leaves slowly open? A fine mechanism is within everything there" ("Siehst Du, wie die saphirblauen großen Blätter langsam sich aufklappen? Eine feine Mechanik steckt da überall darin," 503). One of the more imaginative uses of technology is Weller's attempt to represent the spirit world through Geissler tubes (506). The glowing, glass tubes were named after the German, Heinrich Geißler, who perfected

---

86 "Ein Garten, dessen Teile verstellbar sind. Transportable Hecken, transportable Terrassen. Und besonders.: Transportable beete. Beleuchtung abends durch Glasplatten, die von unten aus erleuchtet werden An Ketten hängende Blumenkorbguirlanden. Riesige Mastbäume mit blühenden Blumen in Erdkörben, die beweglich sind—rauf und runter zu ziehen—und sich auch um den Mastbaum langsam drehen können. Die Blumen müssen aus den Körben lang heraushängen. Verstellbare Blumendrahtwände. Wände zum Schutz gegen den Wind. Das Bewegliche muss im Garten die Oberhand gewinnen: Pflanzen auf grossen Gestellen, die gefahren werden können mit Perpeh. Beweglich Beleuchtung. Schwimmende Beete in den Teichen. Automatische bewegte grosse Fächer mit glänzenden Glasstücken."

the vacuum-pump in 1855 that stabilized the air pressure within the glass tubes (Dörfel 3). The colourful, vibrating light emitted from the Geissler tubes and the many subsequent permutations is a result of gas that has been ionized by a current of electricity (Figure 2). Scheerbart would have seen these tubes with their colourful vibrating light decorating window shops or sold as toys in Berlin (Müller, 215, note 15). In "Flora Mohr," the "phosphorescent flowers" ("phospherescierende Blumen") form one of the five floors in Weller's tower panorama and rotate around the central viewing area. Their phosphorescent glow is emphasized by complete darkness, causing Münchhausen to compare them to "apparitions" ("Geisterblumen," 506). The combination of light, electricity and colour is Weller's attempt to depict the spirit of the flowers: "I just wanted [...] to portray the soul of the flowers—I wanted to offer spirit-flowers" ("Ich wollte mal... die Seelen der Blumen zur Darstellung bringen—ich wollte Geisterblumen bieten." (506). The word "Geist," translated here into spirit and apparitions, is used several times to refer to the phosphorescent flowers, framing light, colour and electricity as a means to depict the immaterial like the phantasmagoria and spirit photography of the 19th century. For Scheerbart as for Weller, vibrating, coloured light becomes synonymous with spirit and forms a part of Scheerbart's technological utopia of a dematerialized world.[87]

The celebratory role of technology in "Flora Mohr" seems at first to be singularly modern, yet its use to animate lifeless material and Scheerbart's scepticism, expressed in his articles, give his texts a particularly Romantic feel.[88] As the scholar of Romanticism John Tresch argues, technology opened up in the arts the possibility of animating the non-living world:

87 When referring to the inhabitants of the asteriod in Leseabéndio as "without character," Kai Pfankuch comes to the conclusion that Scheerbart's literary world is "insubstantial" in the sense of being immaterial (142).

88 The Romantic elements of Scheerbart's fiction may not be coincidental since Scheerbart proclaimed his intention to bring about a new Romanticism in his "Autobiographie":

"Die rasenden Anstrengungen, die ich trotzdem gemacht habe, diese Zeit des Sozialismus, des Militarismus und der Technik zu meinem fabelhaften und sehr religiösen Leben in Beziehung zu bringen, füllen mein sogenanntes Menschenleben aus und brachten meine Bücher hervor, die das Schwervereinbare doch immer wieder vereinen wollen—die eine trockne und fürs Massenhafte interessierte Zeit zu »neuer« Romantik und zu »neuem« Pietismus langsam hinziehen wollen" (329). ("The racing efforts, which I have made in spite of that, this time of socialism, militarism, and technology to bring together with my fanciful and very regious life, fill up my so-called human life and brought forward my books, that intend to unit that which is difficult to unit—the dry and for the masses time to a new Romanticism and to a new piety.").

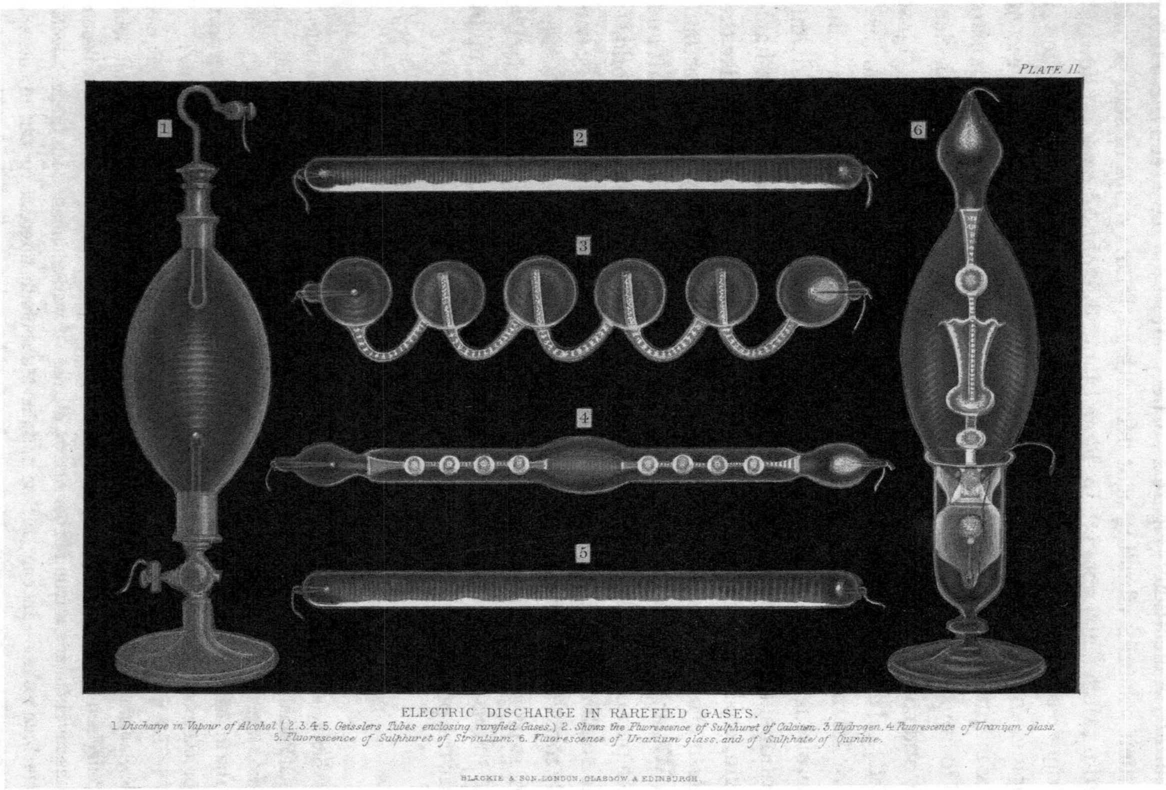

FIGURE 2 *Electric Discharge in Rarefied Gases from M. Rapine* Elementary Treatise on Natural Philosophy, Part 3 Electricity and Magnetism. *(1896) Augustin Private Deschanel. Print. Illustration reproduction courtesy of the* Toronto Public Library.

> The fantastic mode in the arts took shape at a moment when the limits of the possible were being stretched; its scenes of animated matter, vibratory communication, lifelike machines, and eerie metamorphoses implicitly and explicitly referenced the scientific and technological transformations of its time. Rather than read the fantastic as a refusal of positive facts, it participated, along with the nineteenth century's confident new sciences, in a dialectic of doubt and certainty. (126)

Similar to these 19th century writers, Scheerbart expresses both doubt and certainty as to whether discoveries in technology and sciences would have positive impacts for society. He is well-known for his article, "Die Entwicklung des Luftmilitarismus und die Auflösung der europäischen Land-Heere, Festungen und Seeflotten" ("The Development of the Air Militarism and the Dissolution of the European Land Army, Fortresses and Sea-fleets"), on the consequences of developing a military force in the air, foreseeing the impact of air planes on casualties in the successive world wars.[89] In Scheerbart's fictional works and especially "Flora Mohr," he takes a more celebratory note, choosing to imagine the possibilities for technology and science to improve human nature.[90]

Scheerbart's desire to replace biological with artificial reproduction in "Flora Mohr" is a thread common to many Romantic narratives, including Mary Shelley's *Frankenstein* (1817).[91] Weller appears as a reinvention of Mary Shelley's Dr. Frankenstein, replacing biological creation with the artificial

---

89 In Scheerbart scholarship, Scheerbart is often credited singularily prophesying the consequences of using air planes in war. Franz Rottensteiner's afterword to Scheerbart's Rakkóx from 1976 makes clear that such ideas were virulent at the time. He writes: "Aber so neu war diese Sache damals gar nicht mehr, auch wenn es sie in der Wirklichkeit noch nicht geben mochte und auch das Wort nicht; die populäre utopische Literatur der Zeit war schon voll davon." ("But this idea was not any longer so new; even though it may not yet have been available in reality and the term also, the popular utopian literature of the time was full of it," 89).

90 In Münchhausen und Clarissa, he speaks of changing "completely stupid people" ("ganz stupiden Leute") into "moved, sensible artistic natures" ("bewegliche sensible Künstlernaturen") through moving architecture (35). Advanced mechanical devices underlying the aesthetic effect are implied. Elsewhere, Scheerbart more explicitly celebrates the technology underlying his inventions.

91 Movement or metamorphosis is brought to the forefront when the original subtitle of Frankenstein, "A Modern Prometheus", is included. It would be fruitful to compare E. T. A. Hoffmann's Der Sandmann with Scheerbart's "Flora Mohr." While in tone they are vastly different, both consider how inanimate material can be perceived as alive. In Der Sandmann, it is Olimpia who is misapprehended as a living woman through her movement and special glasses.

animation of lifeless materials, glass and iron, and breathing a soul into them with the help of electricity (the Geissler tubes) in addition to mechanical devices (509).[92] But where Frankenstein has created a monster in excess of the human and autonomous, Weller has created beautiful glass gardens in excess of nature—larger, brighter and more beautiful—their movements controlled by Weller through a hidden mechanical apparatus. Scheerbart's utopian vision of artificial creation is missing the ominous tones of Shelley's, whose novel ends with the death of Dr. Frankenstein and his progeny. Instead, "Flora Mohr" ends with a harmonious celebration of technology and nature: "He drove now very fast, that the waves of the great ocean sprayed high from underneath the rubber tires of the baronial automobile. And many white seagulls flew by overhead" (519).[93] Like a high-speed photograph, the last image reflects a perception of a world in motion and life as motion common to both Romantic literature and to Modernism.[94]

The modernist aspect of Weller's plant compositions is apparent through their immateriality and dynamism, which motivates in part the choice of plants over other beings. Scheerbart's vision of a dematerialized world correlates with his self-identification as "anti-erotic" ("Antierotiker"), and his rejection of sexuality (Rausch 616).[95] Scheerbart often derided his friend, author Richard Dehmel, for eroticism in his writing, calling him in a letter from December 1896, a "chief sexualist" ("Sexualisten-Häuptling," "To Richard Dehmel," 36). In his fiction, he privileges spiritual and artistic living over physical desires and substantial bodies. In *Münchhausen und Clarissa*, for example, Clarissa is praised for her desire to dedicate herself to the higher pursuit of art instead of having children.[96] The desire to be married is labelled as particular to women

---

92 Weller says he tried to breath a soul into his glass flowers ("meinen Glasblumen eine sogenannte Seele einzuhauchen").

93 "Der fuhr nun sehr schnell, daß die Wogen des Großen Ozeans hoch aufspritzten unter den Gummirädern des freiherrlichen Automobils. Und viele weiße Möwen flogen vorüber."

94 Dr. Frankenstein not only animated his creation with electricity but his travels throughout throughout the world reflect a newly mobile class of adventurers, whose movements reflect a changing relationship to space and time.

95 In her afterword to *70 Trillionen Weltgrüsse*, Mechthild Rausch identifies Scheerbart's rejection of sexuality (616).

96 Clarissa places herself in opposition to the many who conflate creation with reproduction, insisting instead that a true creative person would overcome this push to reproduce: "die ganze Fortpflanzungsgeschichte ist auf der Erdrinde nur dazu da, daß einzelne schöpferische Menschen die Geschichte überwinden lernen—und bei diesem Überwinden das Schaffen lernen" ("the whole story of reproduction is only there on the surface of the

and represented in "Flora Mohr" by Weller's niece, whose dependence upon Weller's charity to marry her metal worker is manipulated by him to sell his artwork. He uses her steadfast belief in honesty, practicality, the beauty of natural plants and flowers and representational art to form a dark contrast to his glass compositions of light as is indicated by her name Flora Mohr (dark flower).[97] By representing darkness, everything she stands for—established art, the natural beauty of flowers and plants, and physical desires—becomes moralized. In contrast, the immaterial glass compositions become the utopian impulse away from the world of darkness.

The opposition of the material dark world with the light world of glass also corresponds to a physical difference between plants, humans and animals. Plants in "Flora Mohr" are framed as less substantial in their natural form than animals and humans. When responding to Flora's criticism that his flowers are "bloodless spectres" ("blutlose Schemen"), "a shadow realm" ("ein Schattenreich"), and "missing both flesh and blood" ("da fehlt überall das Fleisch und Blut"). Weller points out to Münchhausen that roses and violets are not made of flesh and blood and that Flora's words stem from her desire to be married (503). For Weller as for Scheerbart, there is an underlying distaste for physical desire as represented in the reference to blood and flesh. Plants, missing both flesh and blood, understandably appeal to both as a way to bypass the physical world in an effort to embody the spiritual. In the natural world, plants are already perceived as physically closer to crystal and glass as fleshless and bloodless, and in the hierarchy of being form a link between inanimate minerals and crystals, and animate animals and humans. Weller pushes his glass compositions away from the material world and further into the spiritual by calling them his "glass dreams" ("Glasträume"). Furthermore, the relationship of plants to light in the natural world gives them a certain proximity to Scheerbart's symbol for spirit, light.

The generative process of dreaming reflects two interconnected and fundamental principles in Scheerbart's vision of art and literature, his repeated

---

earth, so that singular creative people learn how to overcome this history—and by overcoming it learn to create") (60). Both Clarissa and Münchhausen belittle those who desire to have a family and hope to have "schöne Gesellschafts revolution" ("beautiful revolution of society"), meaning art would be the point of revolution rather than violence (64).

97 The novella "Flora Mohr" has not been referred to by many scholars. In one of the few interpretations, Heinz Boewe reads Flora as an ironization of Scheerbart's bourgeoisie audience, who would read his narratives as mere "Spielerei" (shenanigans) (30). It would be interesting to further explore the role of Flora in this story as a framing device that offsets the utopian vision of art, part of which involves Scheerbart's satirical image of his opponents: specific audiences, art movements and materialism.

privileging of originality over mere imitation and the difficulty in communicating it. Descriptions of the fantastical plants are frequently accompanied by statements declaring Weller's desire not to imitate natural forms: "When I speak however of lilies, so I use only a word that should give the fantasy a direction; Weller absolutely did not want to offer an imitation of the natural with his glass flowers—nothing was farther away from him than this" (492).[98] The natural world and the language that both describes and determines it is set in opposition to the original, fantastical flowers which cannot be fully explained by the words available. In a discussion between Münchhausen and Clarissa while sitting in a room filled with orchids, originality becomes linked to an excess, an ungraspable that is unsayable: "Naturally, no one understood the main thing. I tried to explain to the best of my ability. Again and again, I said to them that Weller's glass flowers are absolutely not the usual flowers—that they are different than everything that we knew before—that they offer even more than any orchid" (511).[99] This comparison of fantastical plants to orchids is taken from a description of the Australian artist in *Münchhausen und Clarissa*: "[…] the Australian artist has naturally put the biggest effort in all his compositions to offer a deviation from everything earthly. It would be mistaken to talk about orchids, when one actually attempts to trump the orchids."[100] The reference to orchids is definitely not arbitrary. At the time of Scheerbart wrote "Flora Mohr," the orchid was considered an unusual and fantastical flower, inspiring the story "Bologneser Tränen" of Meyrink's short story collection, *Orchideen: Sonderbare Geschichten* (*Orchids: Strange Stories*, 1905) and a drawing by Ernst Haeckel in *Kunstformen der Natur* (*Art Forms of Nature*, 1904). For the Japanese audience in "Flora Mohr," the limits of their imagination stretches to the most fantastical flower in the natural world while misapprehending Weller's originality. For Scheerbart's readers, there is a cautionary note in the Japanese audience's misunderstanding of Weller's flowers, revealing the limits to the short story in describing Weller's new fantastical

98 "Wenn ich aber von Lilien spreche, so bediene ich mich da nur eines Wortes, das ungefähr der Phantasie eine Richtung geben soll; Weller wollte keineswegs mit seinen Glasblumen eine Nachamung der natürlichen bieten—nichts lag ihm ferner als dieses."

99 "Die Hauptsache hat natürlich Niemand begriffen. Ich habs ihnen nach Kräften klargemacht. Immer wieder hab ich ihnen gesagt, daß die Glasblumen Wellers gar keine gewöhnlichen Blumen sind—daß sie anders sind als alles, was wir bisher kannten—daß sie noch mehr bieten wollen als alle Orchideen."

100 "[…] der australische Künstler sich natürlich überall bei seinen Kompositionen die größte Mühe gab, von allen irdischen Abweichendes zu bieten. Es wäre also verfehlt, von Orchideen zu reden; man suchte eben die Orchideen zu übertrumpfen."

flowers. As with Scheerbart's many contemporaries in Symbolism, language was viewed as inadequate to describe the experience of life, and of living art.[101]

Weller's opposition to imitating natural flowers should be read as a way of distancing his flower compositions from the existing art conventions, especially Naturalism and Impressionism.[102] In a self-advertisement for *Münchhausen und Clarissa*, Scheerbart described it as, "only the program for the new artistic goals" ("nur das Programm für die neuen Kunstziele"), a "new art" ("neue Kunst"), and distinct from Naturalism and Impressionism (346, 347). In "Flora Mohr," Weller clarifies his position to representational art when he complains of the Viennese artist, Hans Makart's, reputation as a "Flower-ensouler" ("Blumenbeseeler") even though he painted "only natural flowers" ("nur natürliche Blumen," 509). Known for his intoxicating use of colour in his paintings and his flamboyant decoration, especially the Makart bouquet (Figure 3), the celebrity artist often chose historical subjects for his art and greatly influenced Gustav Klimt's early style. Klimt's later deviation from the Makart style of painting and Makart's own popularity to the point of celebrity contextualizes Makart as perceived to be conventional and unoriginal. The example of Makart also suggests that Weller and consequently Scheerbart wanted to capture the elusive spirit of flowers rather than their physical, material appearance in the flower compositions. Instead of measuring the worth of an artwork or piece of writing through faithfulness to the original, meaning both the object in nature such as an orchid and the established art

101 The Symbolist, Maurice Maeterlinck, in his book The Treasure of the Humble describes the deception of language: "How strangely do we diminish a thing as soon as we try to express it in words! We believe we have dived down to the most unfathomable depths, and when we reappear on the surface, the drop of water that glistens on our trembling finger-tips no longer resembles the sea from which it came" (61).

102 Scheerbart directly criticizes Naturalism as unoriginal in his article, "Hat die Ornamentalkunst jemals nach Originalität gestrebt?": Auch die Neigung zum Naturalismus hat wenig mit dem Streben nach Originalität gemein. Es werden nur die alten symbolischen Blumen—Rosen, Lilien etc. naturalistisch nachgebildet. Die Früchte, Schädel Vögel etc. sind ebenfalls auf eine Verwerthung der alten Ornamentmotive zurückzuführen. Der Nachweis dafür ist an so vielen Beispielen möglich, dass man in den naturalistischen Formen nicht ein prinzipielles Eingehen auf die Naturformen erblicken kann" (102). ("Has the ornamental art ever strived for originality? The leaning to Naturalism also has little in common with the striving for originality. Only the old symbolic flowers—roses, lilies etc are being naturalistically imitated. The fruits, skulls, birds etc can be similarily linked back to a devaluing of the ornamental motif. The proof for that can be seen in so many examples that in the naturalistic forms, a basic link to the forms in nature cannot be seen.")

So liess man Blumen sprechen

Text und Bild aus dem Jahre 1884

Das also war das Schmuckstück in den Salons unserer Eltern: der Makartstrauß, ein Staubfänger aus künstlichen Palmenwedeln, Rohrkolben und Gräsern. Er galt damals als Sinnbild der „Raumkunst". Wie man ihn seinerzeit begrüßte, lesen Sie in nebenstehenden Worten

Wohl selten hat sich ein Zimmerschmuck so rasch und siegreich Bahn gebrochen, als der nach dem berühmten Maler getaufte Makartstrauß.

Man hat einsehen gelernt, daß gerade das Farblose, „das Tote", wie ein Vorurteil es nannte, bestimmt ist, dem Bouquet die Berechtigung zu verleihen, sich ebenbürtig feinen Luxusgegenständen anzureihen.

Wie von der Rücksichtnahme der Farbenwirkung abgesehen werden kann, so bindet sich die Verwendung von Makartbouquets also auch nicht an Räume von einem bestimmten Umfang. Meterhohe Kolossalsträuße zieren die Ecken und Nischen der Ballsäle, die Vestibüle, Foyers. Die Mittelgröße findet ihren Platz in Speisesälen, Teezimmern, Salons und Esszimmern, während selbst in das kleinste, traulichste Boudoir der Makartstrauß in Miniaturform Eingang zu finden weiß. Was die Herkunft des Materials anbelangt, so übernimmt der Strauß gleichsam die Vertretung der Pflanzenwelt aller Himmelsstriche. Die weißgebleichten, fiederreichen Palmwedel entstammen Ägypten, die silberglänzenden cremefarbigen Pampaswedel nennen die Prärien Kaliforniens ihre Heimat, Algier, Italien und Südamerika senden diese, Ungarn, Rußland jene Gräser und auch Deutschland ist mit verschiedenen kleinen Zittergräsern vertreten.

340

FIGURE 3 *Makartstrauß from Scherl's Magazin. 6 June 1933: 340. Illustrierte Magazine. Web. 10 Jan 2014.*

conventions, Scheerbart advocated for literature and visual art that sought to be original and spiritual.

However, while Scheerbart views representational art as unoriginal, this does not translate into antipathy for the natural world. In her book on *Münchhausen und Clarissa*, Beatrice Rolli asserts that Scheerbart views art as a way to enhance the magnificence already present in nature (16). Her assessment is confirmed by a contemporary of Scheerbart's, Max Creutz, who attested to Scheerbart's affection for nature, when he wrote that nothing would give Scheerbart greater joy than to receive an orchid or another small flower as a gift (369).[103] Karl Hans Strobl in a small article about Scheerbart described him as "a person of love, of love of nature and all of her creatures" ("Scheerbart ist ein Mensch der Liebe, der Liebe zur Natur und zu allen ihren Geschöpfen," 370). A similar sentiment is repeated by Weller in "Flora Mohr" in defence of his glass flowers: "I am of course also a great friend of natural flowers" ("Ich bin ja auch ein großer Freund der natürlichen Blumen," 506). The reaction of the Japanese audience to Weller's flower compositions also frames the relationship of the natural flowers to the artificial ones as complementary rather than oppositional: "This Flora Mohr is incomprehensible to us. We understand Mr. Weller's palace as the highest glorification of the flower world. We couldn't see it otherwise" (510).[104] For Weller as for Scheerbart, creating artificial glass flowers does not mean that he is against everything natural, but signals a complex relationship to nature that venerates the natural through expressing the creative force alive in the entire universe from the flower to the cosmos.

Weller is not only venerating the dynamic force he believes drives creativity; he also pays tribute to one of the foremost observers of plant movements from the Romantic nature philosophers, Johann von Goethe. The reoccurring spiral forms in many of Weller's flowers recall Goethe's observations of plant growth in his essay, "Der Versuch die Metamorphose der Pflanzen zu erklären" ("Metamorphosis of Plants," 1790) and his subsequent didactic poem "Die Metamorphose der Pflanzen" ("The Metamorphosis of Plants," 1798).[105] Weller's references to Goethe are framed as a tribute in a circular foyer that is

---

103 Max Creutz writes: "Und man konnte dem 'armen' Paul Scheerbart keiner größere Freude machen, als wenn man ihm eine Orchidee oder auch eine kleinebescheidene Blume schenkte" (369).

104 "Diese Flora Mohr ist uns ganz unverständlich. Wir nehmen Mr. Weller's Paläste als höchste Verherrlichung der Blumenwelt hin. Wir können garnicht anders."

105 In a later essay from 1831 ("Über die Spiraltendenz der Vegetation," "On the Spiral Tendency of Vegetation"), Goethe directly addresses spiral movement in plants and attributes spiral

staged as a sacred space, an "age-old church" ("uralte Kirche"), and a museum, an "age-old rotunda" ("eine uralte Rotunde") where the room must be protected from wear through slippers. Given Scheerbart's fascination with World Fairs, the rotunda also acts as an exhibition hall like the Viennese rotunda built for the 1873 World Fair (492–93).[106] The space is dominated by three oversized lily plants in the centre of the room with "snow white stems" ("schneeweiße Stengel") "lemon-yellow horn-shaped blossoms" ("citronengelbe dütenförmige Blüten"), and "long iridescent filaments like mother-of-pearl [...] in spiral forms" ("lange perlmutterartig schillernde Staubfäden [...] in Spiralformen," 492). Unlike the plants in most other rooms, the glass flowers appear still and inanimate, not yet speeded up to human time, at the pace that Goethe and his contemporaries would have encountered them in visual art and in the natural world. The stillness of the glass flowers and the reference to movement in the spiral forms honour a past in which plant movement was not yet visible to the naked eye but still seen and expressed through poetic language. For Goethe, poetic language offered a temporal medium to present development and life cycles. Goethe's vision of the spiral growth of plants and his acceleration of a plant's life, making visible its slow movements, are exemplified in his poem, "Die Metamorphose der Pflanze" (1798). He writes: "In careful number or in wild profusion / Lesser leaf brethren *circle here the core.* / The crowded guardian chalice clasps the stem, / Soon to release the blazing topmost crown / So nature glories in her highest growth" (emphasis mine).[107] Goethe's dynamic plant is echoed in Weller's rotunda through the adjectives of motion describing the glass lilies, "turned" ("gedreht") and "wound" ("gewunden"). Similar to Goethe, Scheerbart chose the medium of letters to express the foundational aspect of life as temporal, yet his vision of glass architecture and plants, and the limits he saw in language express a dissatisfaction with what he and his contemporaries understood as the crystallizing properties of language. The limits of language to express the knowledge of the natural world resonates with Goethe's view of taxonomy. He positions his vision of a dynamic plant against

growth to the feminine half and linear growth to the masculine, resulting in an androgynous plant (np).

106 It is fair to assume that Scheerbart would have known about the Viennese rotunda given his apparent fascination with world fairs. He uses the visionary concept of demonstrating the technological prowess as the staging ground for his vision of art and architecture in Münchhausen und Clarissa. Many of the media, such as the phantasmagoria and the panorama were also part of the world fair.

107 "Rings im Kreise stellet sich nun, gezählet und ohne / Zahl, das kleinere Blatt neben dem ähnlichen hin. / Um die Achse gedrängt, entscheidet der bergende Kelch sich, / der zur höchsten Gestalt farbige Kronen entläßt."

the taxonomical method of understanding the world, suggesting that naming and classifying limits knowledge of the natural world rather than expanding it. Scheerbart also echoes this perspective and takes it a step further, at once, paying tribute to Goethe in the spiral forms and recognizing that existing knowledge of the natural world, the names and categories, limits the understanding of the generative and creative power in nature.

Recognizing the limits of language, especially the word "soul," and examining those limits was part of Gustav Fechner's pan-psychic project. Gustav Fecher, one of the greatest influences on Scheerbart, was in turn greatly influenced by Goethe and other Romantic philosophers of nature. Writing at the mid-19th century, Fechner sought to bring together the hard science of empirical research and metaphysical philosophy to create a holistic perspective that would unite the materialism of science with the anti-materialism of spirituality. Fechner's impact on Scheerbart's concept of a "world soul" ("Weltseele") has been recognized and often discussed in reference to *Lesabéndio* and other depictions of comets and stars. In his examination of science and fantasy in Scheerbart's *Lesabéndio*, Müzeyyen Ege sees the central worth of Fechner's world view on Scheerbart as the "cosmo-fantastic" ("'kosmophantastischen' Entwürfen") design of *Lesabéndio*, where the comets and stars possess a soul and their own will (77).[108] Ege concludes that Scheerbart found in Fechner's cosmo-psychology a "'fantastical science,' in which the seemingly 'incompatible' values of rationality and mysticism can be united. This means for him first of all an expansion of the reality concept, in which the fantastical is closely linked with reality rather than standing in opposition to it" (85).[109] Scheerbart did directly refer to Fechner's influence on his view of the cosmos in his article "Das Ende des Individualismus" ("The End of Individualism," 1895). He claims that Fechner's *Zend Avesta oder Über die Dinge des Himmels und des Jenseits, Vom Standpunkt der Naturbetrachtung* (1851) had developed "the basic principles of the cosmo-psychology" ("die Grundprinzipien der Kosmospsychologie") defined as a "new, highly fantastic science" ("neue, höchst 'phantastische'

108 Fechner is commonly cited as exercising a profound influence on Scheerbart. Like Ege, Parsch's article on science fiction and Lesabéndio notes Fechner's influence on Scheerbart's cosmo-psychology. Although Parsch does not connect Fechner and Scheerbart's portrayal of plants, he does include a little known connection between a satirical philosophical essay by Fechner on angel bodies as spheres and Lesa's meditations on the "intoxicating" movements of galaxies (213).

109 "In der Kosmopsychologie Fechners schließlich findet Scheerbart eine 'phantastische Wissenschaft, in der das 'Schwervereinbare', Rationalität und Mystik, vereint werden können. Dies bedeutet für ihn in erster Linie ein Ausweitung des Wirklichkeitbegriffs, in der das Phantastische mit der Realität eng verbunden ist und keine Opposition darstellt" (85).

Wissenschaft," GW 256). *Zend Avesta* was framed by Fechner as an expansion of the idea of a plant soul outlined in his earlier work *Nanna, oder über das Seelenleben der Pflanzen* (1848) to the possibility of a cosmic soul.[110] Although Ege mentions *Nanna*, he does not link Scheerbart's portrayal of plants as ensouled, living beings with Fechner. In *Nanna*, Fechner premises the possibility of a plant soul on what he perceives to be a mistaken limitation on who can possess a soul, instead believing that individual souls are "particular manifestations of a general soul" ("individuelle Seelen als besonder Ausgeburte der allgemeinen Beseelung," 33). Scheerbart adopted this idea, calling it a "world soul," claiming that "we humans do not actually think at all, but rather the Earth thinks only through us" ("wir Menschen eigentlich überhaupt nicht denken, sondern daß nur die Erde durch uns denkt," 255). In reference to the Earth, Scheerbart further exclaims that, "The stars think!" ("Die Sterne denken!"), thus attributing sentience to the planets (255). As with Fechner, Scheerbart believed that all earthly and astral beings are different manifestations of the world soul, leaving the pinnacle role, conscious thinking, to humans. The impact of Scheerbart's admiration for Fechner's version of pantheism on "Flora Mohr" is essential for understanding the philosophical underpinnings of Scheerbart's dynamic plants beyond their resemblance to the serpentine lines in Jugendstil art.[111]

Apart from "Das Ende des Individualismus," Scheerbart directly addressed the possibility of consciousness and life in the cosmos in two other articles written around the same time as "Flora Mohr." In "Sternschnuppen und Kometen" ("Falling Stars and Comets," 1909) and "Sind die Kometen lebendige Wesen?" ("Are the Comets Living Beings?," 1910), Scheerbart uses an analogy between plants and comets to argue that the movement of comets is evidence that they are living and possibly sentient, just as plant movement

110 Gustav Fechner states directly in his introduction that Zend-Avesta is a contuance of the project he started in Nanna: "Eine frühere Schrift, Nanna, kann insofern als Vorläuferin der jetztigen gelten, als dort wie hier versucht wird, das Gebiet der individuellen Beseelung über die gewöhnlich angehommenen Gränzen hinaus zu erweitern; dort aber in abwärts gehender, hier in aufwärts gehender Richtung" (IV) (An earlier writing, Nanna, can be considered as a predecessor to the current one, for there as here it is being attempted to increase the area of the individual soul over the accepted boundaries; but there the direction looks downward and here it looks upward).

111 Scheerbart wrote of life in the cosmos in several of his fictional texts including, Lesabéndio, ein Asteroïden-Roman (1913), Die Seeschlange (1901), Liwûna und Kaidôh. Ein Seelenroman (1902), Kometentanz. Astrale Pantomime in zwei Aufzüge (1903) Astrale Novelletten (1912).

indicates life and a soul.[112] In "Sternschnuppen und Kometen," he reads the path of meteorites as a sign that they are "intelligent" ("vernunftbegabt," 376). In "Sind die Kometen lebendige Wesen?," he, at first, cautiously compares orchids and comets: "We need not at first think of rational, thinking beings—we can compare of course the comets with orchids. That this is a real 'life' cannot be denied; every researcher of nature would agree without a second thought" (449).[113] Scheerbart continues with the analogy of plant life, using the possibility of plant intelligence to argue for the possibility that comets are "'intelligent' beings" ("'vernünftige' Wesen") suggesting that the simple plant soul may be equivalent to the comet soul (453). Just as Gustav Fechner began with the possibility of a plant soul and then extended his definition to include the possibility of a cosmic psyche, Scheerbart follows the same line of reasoning, extrapolating from life on earth to the cosmos. At the end of the article, Scheerbart concludes that the way has been made "for a new astral world view, in which the comets and planets—the moon and the suns—lead a free, big, cosmic life, in the face of which we have to bow to in respect (453).[114] For Scheerbart, the analogy of plant life and the cosmos was especially meaningful for the possibility of a new understanding of planetary movement. So, when his character, Weller in "Flora Mohr," says that after seeing a comet appear to merge with the sun, he imagined colossal life in a comet and that this led to the idea that "the comets could be colossal flowers" ("Und ich stellte mir das kolossale Leben in diesem Kometen vor," 515), the resulting glass creation, the comet flowers, should be read as an attempt to create a fictional account of Fechner's idea of a world soul.

---

112 In Münchhausen und Clarissa, he takes the analogy a step further to illustrate the advanced nature of the literature in Australia: "Andrerseits wird auch unsre Erdensonne in die Literatur als selbständig denkendes Lebewesen eingeführt. Daß irgendein Stern ein totes dummes unorganisches Ding sein könnte—daran denkt in Melbourne wahrhaftig Niemand mehr; das wäre ja auch zum Lachen, wenn man das Größere so ohne Weiteres für das Dümmere halten wollte. Die Literatur steht also in Melbourne im innigsten Zusammenhange mit der bildenden Kunst; hier wie dort will man das Neue um jeden Preis. (69).

113 "Wir brauchen dabei zunächst noch nicht an vernünftig denkende Wesen zu denken—wir können ja die Kometen mit Orchideen vergleichen. Daß diesen ein wirkliches »Leben« nicht abzusprechen ist, das wird ja jeder Naturforscher ohne weiteres zugeben."

114 "für eine neue astrale Weltanschauung, in der die Kometen und Planeten—die Monde und die Sonnen—ein freies großes kosmisches Leben führen, vor dem wir uns in Ehrfurcht beugen müssen."

Scheerbart uses "Flora Mohr" to visualize manifestations of the "world soul" in plants, extending it, as well, to astral life. Weller's last and "best" artwork, the comet flowers, forms an analogy between astral life and flowers by bringing them together in one artwork as a way of emphasizing their similarities. He blends the two forms together: "The sun is also made from glass. You see how it always changes the colour. And the flowers surrounding it always receive different tails—by every small movement the tails are changing" (515),[115] and then later on: "[...] that also on the surface of the sun, images of flowers would also be visible" ("[...] daß auch auf der Oberfläche der Sonne Blumengebilde sichbar würden," 516). Weller voices the analogy as above all a similarity of form: "In any case [...] it is not improbable for me that the protuberances of the sun possess a certain flower-like character. One already speaks so often of protuberance forests" ("'Jedenfalls', sagte Weller, 'ist es mir nicht unwahrscheinlich, daß die Protuberanzen der Sonne gewissen blumenartigen Charakter besitzen; man spricht ja schon so oft von Protuberanzenwäldern.'," 516). From the similarity in form and movement, Weller implicitly draws the conclusion that comets and flowers both possess life and a soul in the natural world and in his glass compositions with the ambivalent statement: "There, everything is moving. There, nothing is at all dead" ("Da bewegt sich Alles. Da ist garnichts tot," 515). Movement in the glass compositions was not only a sign of life but of a soul. In *Zend-Avesta*, Fechner refers to the movements of the earth as "soul movement" ("Seelenbewegung") bringing together movement, life and spirituality (105). The analogy of flowers to comets as a way of seeing life in the astral movement also has the opposite effect of causing flowers to be viewed as the foreign and astral life on earth.

Scheerbart's admiration for Fechner's ideas belongs to a surge in interest at the turn of the century in the concept of a world soul. Kurd Lasswitz, another German language pioneer of science fiction, wrote a book examining Fechner's oeuvre, and a novel applying Fechner's ideas to fiction, *Sternentau: Die Pflanze vom Neptunsmond* (*Sun Dew: The Plant from Neptune's Moon*, 1909).[116] Another

115 "Die Sonne ist auch aus Glas. Sie sehen, wie sie sich immerzu in der Farbe verändert. Und die Blumen ringsum bekommen immer wieder andere Schweife—bei jeder kleinen Bewegung verändern sich die Schweife."

116 The name of the plant invokes a connection between astral life and plants. As with Scheerbart, Laßwitz uses fiction to demonstrate Fechner's concept of a world soul. In a conversation between a human and a plant, they discuss the different manifestations of the world soul: "Wir Menschen wollen auch teilnehmen an dem Ganzen des Erdlebens. Auch wir begreifen mehr und mehr, daß alle Lebewesen nur Organe sind des Planeten und daß unser Bewußtsein in seiner großen Einheit zusammenhängt. Nur geht unser Weg von der Einzelseele zur Allseele, der eure aber umgekehrt von der Dauerseele zur

contemporary of Scheerbart's, Raoul Francé, also brings together plant movement and spiritual energy. He writes: "The plant possesses everything that distinguishes a living creature—movement, sensation, the most violent reaction against abuse, and the most ardent gratitude for favours" (20). He also connects movement and plants to a creative force in the universe, concluding that:

> [...] life is a special force, standing on an equality with the other forces of nature, which transforms the raw material into something whose final form is unfortunately still concealed from us by the clouds of our own limitations. In this unfortunate condition of uncertainty, the one firm point to which we can cling is the feeling of complete inner unity with the creative and transforming forces of nature.
>
> *The Germ of Mind in Plants*, 126

The belief in a world soul is common to all three writers and points to a need at the turn of the century to balance materialism and a taxonomical understanding of the natural world with a perspective that allows for a spirituality based on movement.

Movement is combined with colours and forms in "Flora Mohr" to create a synesthetic experience that invokes plant spirit and life. The phrase, "colours and forms," is used repeatedly throughout the novella to describe Weller's glass flowers as living and possessing a soul. Coloured glass combined with light draws upon the sacred iconography of the church as a visual reminder of the heavenly spirit. In Scheerbart's manifesto on glass architecture, he explicitly places his use of coloured glass and light within a spiritual tradition from the ancient Babylonian temples and mythical Alhambra palace to Gothic churches, and adapts this sacred and spiritual purpose for widespread and common use (130). According to scholar Andre Schuelze, Scheerbart also draws upon theosophical thought, when he connects light with a "visualization of the immaterial, metaphysical powers" ("Sinnbild der immateriellen, metaphysischen Kräfte," 41). As in both spiritual tradition and theosophical thought, Münchhausen encourages the interpretation of the "colours and forms" in Weller's glass composition as the visualization of a life force: "And then Weller pushed a button, and there were again other buds there—in

---

Einzelseele (Pflanzenrede)." ("We humans also want to participate in the entirety of the earth's life. We also understand more and more that all living beings are only organs of the planets and that our consciousness is connected into its greater unity. Only, our way goes from the single soul to all soul, and that of the plant is the opposite from the everlasting soul to the single soul (plant talk).")

other forms and colours. 'Now,' called William laughing, 'is this spring glory living—or is it not living?' " (507).[117] For Weller, the transformations of colours and forms suggest life. They also point to a modern adaptation of the static church windows, reflecting the meaning of motion for Scheerbart and his contemporaries.

Even though the kaleidoscopic colours and forms are a visualization of the creative and generative life force, they do not represent the essence of the natural world. Flora's objections to Weller's compositions reflect a central epistemological problem. How to represent the subjective experience of living beings, plants, animals (and even comets, if we are to subscribe to Scheerbart's belief that they are living beings), for which we have no experience and no material basis. For the materialist, Flora, non-representational art and the medium glass are empty of meaning, visualizing neither life nor the soul. She asks Münchhausen and Weller: "Can you deny that they're dead? And—is it not always the same, what one sees here? Always again only colours! And always again only forms?" (499).[118] She repeats herself later, shifting her argument slightly, from lifeless to soulless: "There are always again only colours and forms—nothing more. The soul is missing" ("'Da sind immer wieder nur Farben und Formen—wieder nichts Es fehlt die Seele!' ", 507). In response to her objections, Weller, then, clarifies the position of his artwork to nature as free from claims of truth that maintain they represent the heart of nature: "The Flora is right in a manner of speaking. I admit that only colours and forms ever appear. But—is it not a little bit presumptuous, when one wants to discover immediately the quintessence of nature" (508).[119] Scheerbart claims here and elsewhere that the inner life of animals and plants cannot be known but traces of an inner life can be read in their movement.[120] Flowing colours and forms by being non-referential are another way that the spirit of nature can be invoked without claiming to know it or determine it.

Flora's role as the critical voice reflects a vision of femininity as negative, conservative and bound up with the body. Unlike Flora from the film *Das*

117 "Und dann drückte Weller auf einen Knopf, und es waren wieder andere Knospen da—in anderen Formen und Farben. 'Nun', rief William lachend, 'lebt diese Frühlingspracht—oder lebt sie nicht?"

118 "Können Sie leugnen, daß sie tot sind? Und—ist es nicht wieder dasselbe, was man hier sieht? Immer wieder nur Farben! Und immer wieder nur Formen!" (499).

119 "Die Flora hat ja gewissermaßen Recht. Ich gebe zu, daß immer nur Farben und Formen kommen. Aber—ist es nicht ein bißchen anspruchsvoll, wenn man immer gleich den Kern der Natur entdecken will?"

120 In Scheerbart's article on comets, he argues that the soul of orchids cannot be known but the fact that they exist is understood.

Blumenwunder, discussed in chapter three, and Harda from Kurd Lasswitz's novel *Sternentau*, Flora as a materialist does not reflect a progressive attitude towards women. While the Flora from *Das Blumenwunder* reflects the reframing of plants as active and progressive, Flora represents the conservative attitude that refuses to envision a new future. Likewise, Harda's interest in technology and science stands in opposition to Flora's preference for the natural. Scheerbart's short story is one example where the dynamic vision of plants does not extend to the view of femininity. Instead, Flora is represented as standing still amidst her uncle's shifting kaleidoscope of dynamic flowers.

The kaleidoscopic effect of the constantly shifting colours and forms figures prominently in many of Weller's glass displays as a means of invoking a sense of incompletion familiar to both the Expressionist crystal metaphor and plant growth. In Weller's experiment room, he uses a series of magnifying lenses of differing strengths set in the floor to create a "kaleidoscopic effect of the best sort" ("einen Kaleidoskopeffekt erste Güte") with the moving glass flowers underneath. Münchhausen describes the effect as an overwhelming combination of magical light, colours, and forms ("Lange hielt ich natürlich diesen Farben-, Formen- und Funkenzauber nicht aus," 494). Each combination of the lenses in the floor and the placement of the flowers underneath produces an entirely new visual experience, and simultaneously suggests the potential for an endless array of new combinations. Rosemarie Bletter sees many similarities between Scheerbart's concept of glass architecture and Expressionist crystal metaphor: "... if there is an ideal, it is incompletion and tension: shifting, kaleidoscopic forms are forever moving out of chaos toward a potential perfection, a perfection which is, however, never fully attained" ("The Interpretation of the Glass Dream-Expressionist Architecture," 33). The crystal metaphor of the constantly shifting forms transfers easily to plant form, reflecting on how plant growth is in constant movement towards a 'potential perfection,' to Goethe's primordial plant—always adding new leaves to the old, while never attaining the completed form.

The kaleidoscopic vision of plant movement also resonates with the evocations of plants by one of the greatest influences on Art Nouveau, dancer Loïe Fuller, whose efforts to convey the impression of pure energy in her veil dances relied on a combination of spiritualism and technology. Both Scheerbart and Loïe Fuller blend coloured lights and movement to evoke plant forms without becoming referential. In addition to her famous serpentine, fire, and butterfly dances, Fuller invoked various flowers, including orchids and lilies, in her light and colour dances as an effort to communicate the sensual experience not captured in conventional depictions of flowers (Figures 4 & 5). She also projected microscopic photographs of cancer cells, images of skeletons and the

FIGURE 4 The Dance of the Lily *from Samuel Joshua Beckett. 1900. Illustration reproduction courtesy of* The Metropolitan Museum of Art, *Gilman Collection, Purchase, Mrs. Walter Annenberg and The Annenberg Foundation Gift, 2005 (2005.100.961) Image © The Metropolitan Museum of Art.*

FIGURE 5 La Loïe Fuller *from Jules Chéret.* NYPL. *Web. 27 Feb. 2016. Illustration reproduction courtesy of the New York Public Library.*

surface of the moon onto her luminous garments. Tom Gunning, in an article on Loïe Fuller and Germaine Dulac, draws attention to plants as an inspiration for both the dancer and the filmmaker's experiments with kaleidoscopic visual spectacles. In describing Fuller's flower dances, he writes: "The flowers formed by Fuller's rapidly swirling fabrics could never be static, but rather traced the pathways of motion itself, holding within each of its evanescent forms the imminence of transformation. Anticipating Dulac, Fuller declared, 'I desire to materialize the *insaisissble!* [unsayable]" (115). In her memoir, full of references to the many famous people she encounters, Fuller also writes in a short chapter on her belief that motion and light can convey more fully than language the experience of the senses (Fuller 72). Her attraction to light, colour and movement as the basis of her sensory spectacles echoes Scheerbart's efforts to combine impressions from the senses in a constantly shifting play of light and colour in "Flora Mohr." That plants are one of the fundamental points of inspiration for both Fuller and Scheerbart points to a widespread shift in the perception of plants from inanimate objects to dynamic, living beings at the *fin de siècle*.

Fuller and Scheerbart's choice of ethereal media: light, colour and movement made it a short step from portraying plants as living beings to visualizing the plant soul. It is clear from Fuller's memoirs that she believed she was channelling the soul of nature in her dances. She quotes from an audience member's journal that describes her impressions of Fuller's performance: "Soul of flowers, soul of the sky, soul of flame, Loïe Fuller has given them to us. Words and phrases avail nothing" (264). The audience member frames Fuller's fluid dances as capable of expressing the immaterial and the evanescent, which is beyond the determining properties of language. The same capacity to express the unsayable is attributed to Weller's plant compositions, "what I got to see with two crutches under the arms, doesn't let itself be so easily described. I saw into a colourful, completely marvellous flower and fruit world, and with every slow step, sliding on fur shoes, the colourful image changed underneath us" (494).[121] Created from glass, light and movement, Weller's flowers and plants also reflect the elusive, immateriality attributed to the concept of the soul,

121 "[...] was ich da mit zwei Stricken unter den Armen zu sehen bekam, läßt sich leider nicht so leicht beschreiben. Ich sah in eine bunte, ganz fabelhafte Blumen- und Fruchtwelt hinein, und mit jedem langsamen Schritt, auf Filzschuhen rutschend, veränderte sich das bunte Bild unter uns."

while holding on to a sense of permanence missing from Fuller's evanescent dance and present in photography and film.[122]

As with Fuller, Weller's plant compositions also anticipate film, drawing on a combination of light and motion to evoke plant-like forms, but they also directly prefigure the avant-garde filmmaker, Germaine Dulac's, concept of film as visual music. While Dulac encouraged scientific and documentary style films, her own project strayed from the direct representation of the natural world. She conceived the potential of cinema as, "a visual symphony made of rhythmic images, coordinated and thrown upon the screen exclusively by the perception of the artist" (41). Dulac's visual music resonates with Scheerbart's method for "composing" original art and architecture. In his advertisement for his novel *Münchhausen und Clarissa*, he writes: "I don't want to create through imitation, but rather through free composition of the impressions of the senses that which can be labelled 'new art'" ("Ich will nicht durch Nachbildung, sondern durch freie Komposition der Sinneseindrücke das schaffen, was man als »neue Kunst« bezeichnen kann," 346). In the novel *Münchhausen und Clarissa*, he elaborates on his intention, suggesting that composing music is comparable to assembling image pieces from the natural world:

> The Australian painter believes that he comes much faster behind the essence of nature, when he separates the single pieces of nature from one another and brings them together again afterwards in another art. Creating means for the Australian: New creating! And he can only create something new in his opinion, when he takes apart the available nature images—and with the taken-apart pieces, he creates new—completely new—images. Creating is composing. And one composes in Melbourne not only in music—one composes there in all arts. (80)[123]

---

122 The permanence of glass was very important for Scheerbart. In his manifesto on glass architecture, he writes: "Alle Baumaterialien, die haltbar und in wetterbeständigen Farben zu erhalten sind, haben Existenzberechtigung. Der zerbröckelnde Backstein und das brennbare Holz haben keine Existenzberechtigung" (45). ("All building materials that last in weatherproof colours have a right to exist. The bricks that crumble and the wood that burns have no right to exist.")

123 "Der australische Maler glaubt, daß er hinter das Wesentliche der Natur viel schneller kommt, wenn er die einzelnen Stücke der Natur voneinander trennt und sie nachher wieder in andrer Art zusammenbringt. Schaffen heißt für den Australier: Neues schaffen! Und Neues schaffen kann er nach seiner Meinung nur, wenn er die vorhandenen Naturbilder zerlegt—und mit den zerlegten Stücken neue—ganz neue—Bilder schafft. Schaffen ist eben komponieren. Und man komponiert in Melbourne nicht nur in der Musik—man komponiert dort in allen Künsten."

In "Flora Mohr", Weller's glass flowers and plants are referred to as "compositions" ("Kompositionen," 509), which then explicitly frames the "compositions" as an application of Scheerbart's method of composing visual music. The glass flower and plant compositions not only appear kaleidoscopic but the method of viewing the natural world is in essence also kaleidoscopic.

Scheerbart also shares with Dulac the belief that visual music is best suited for representing the unsayable. As for Scheerbart, plants were a key source of inspiration for Dulac's visual music, appearing in her writings as examples of rhythm and the visual capacity of film to reveal the imperceptible, the ungraspable and the emotional drama inherent in life itself. Her kaleidoscopic combination of light, water, plants and crystals in her film *Arabesque* (1929), "aspire less to present recognizable images than to invoke the *insaisissable* energies behind them, the flow of motion" (Gunning, "Light, Motion, Cinema!," 122).[124] As in Dulac's films, Weller's non-referential compositions have been privileged as particularly capable of communicating the flow of motion, here called "spirit": "Be still, I said, the Nabob and I are not like your lovely Flora; we see alright, what a wealth of spirit stories live inside us and come into us—like music coming into us" (510).[125] Moving visual media are framed as the ideal "language" for depicting the intangible aspect of a plant's life, its temporality or lived experience that contrasts so starkly with a static image of a plant as only material—an object.

Weller's plant compositions also anticipate film in another way through the radical reconfiguration of the panorama. The panorama during the 19th century was a dynamic mixed media that used techniques such as lighting,[126] masking the frame, incorporating sculpture and the 360 degree view to give the illusion of reality (Figure 6). Some installations consisted of a rotating canvas that presented a seated viewer with a series of scenes creating a

124 While I have not been able to view this film, Tom Gunning has provided a detailed description of it in his article on Fuller and Dulac: "The finale of the film presents a time-lapse close-up of a flower as its petals unfurl, followed by crystals in growth, the shot of a twirling motion in a distorted mirror, a kaleidoscopic motion of silver balls. An even closer view of a woman's face appears, her face is juxtaposed with the fully unfurled flower, its white petals echoing her white scarf. In the final shot, the fountain transforms, in a subtle dissolve, into a faceted mirrored ball, which seems to cast both light and fragments of a reflected image of nature back into the camera" (121).

125 "Sei still, sagte ich, der Nabob und ich sind nicht so wie Deine liebe Flora; wir sehen schon, welche Fülle von Stimmungsgeschichten leben und kommen in uns hinein—wie Musik in uns hineinkommt."

126 Daguerre's painted dioramas from the 1820s made use of light to illuminate painted scenes and create a sense of wonder what viewers dubbed the Salle de Miracle.

FIGURE 6
Section of the Rotunda, Leicester Square, In Which is Exhibited the Panorama from Plans, and Views in Perspective, with Descriptions, of Buildings Erected in England and Scotland *by Robert Mitchell. London: Printed by Wilson & Co. for the author, 1801.*

narrative of the event as well as the impression of passing time.[127] Panoramas usually depicted historical or current events or what Alison Griffiths refers to as "reenactments," meaning that the events, "were to be interpreted *as if* the action was happening along an immediate temporal and spatial presence and continuity" (2). Weller's tower panorama shares with the 19th century version this immediacy, but instead of reenactments, it generates a new, non-referential fantasy world that replaces the representational paintings with glass sculptures. Weller also retains many of the characteristic features of the panorama, including the large scale, fixed spectatorship, invocation of presence, and delimitation of boundaries that anticipate film. Furthermore, he enhances the lighting effects, bringing his version of the panorama closer to experimental film, especially to Germaine Dulac's kaleidoscopic plant films rather than to the early nature films (Figures 7 & 8).

As with Dulac, Weller's tower panorama is an attempt to create visual music out of an interplay of light and shadow rather than simply animate the slow-moving plants through time-lapse photography as in the early nature films. The first four floors of the five-storey tower anticipate the reduction of objects to light and shadow in experimental film from the twenties. The viewing experience of his panorama is placed between panorama and film. Weller and his company are seated in plush chairs as if in a cinema, watching the first floor, a frozen winter landscape of ice flowers, move past them as a moving panorama.[128] As common to panorama installations, they enter from the middle of the floor into the second storey, the shadow play, where the spectacle comes to resonate with contemporary film language: "There, we sat just like

127 "The Wreck of the Medusa" was such an example of a Peristrephic panorama. A long strip of canvas that contained painted scenes was unrolled in front of a visitor.

128 Scheerbart is playing with the word Eisblumen which means frost but can be broken down into ice-flowers. The icy flowers resemble frost, the genesis of their name.

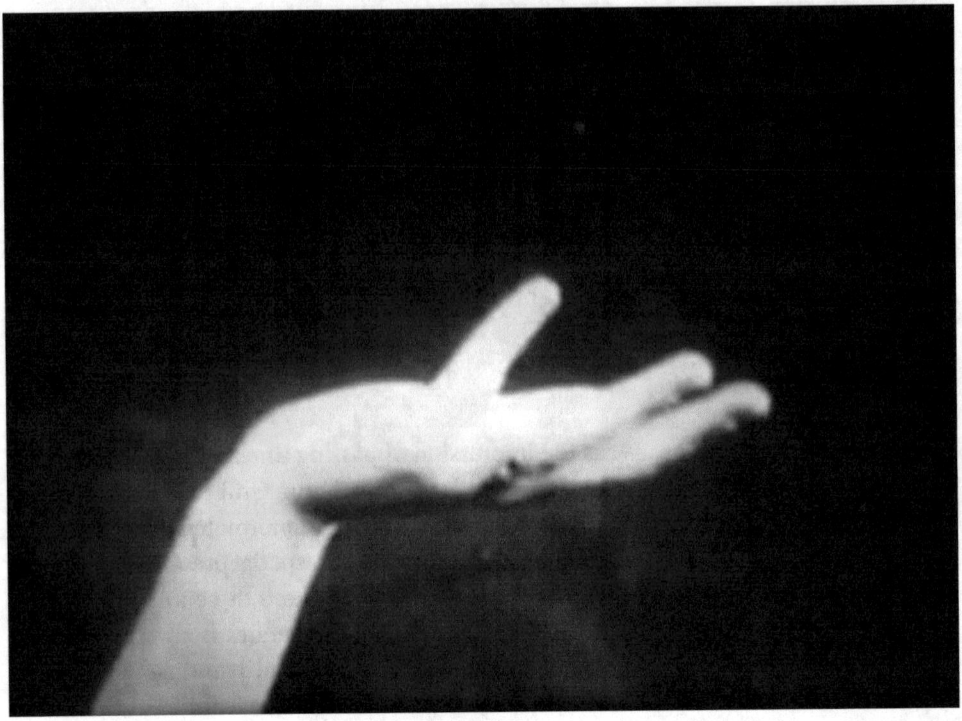

FIGURE 7 *Hands from* Thèmes et Variations. *Scene still. Dir. Germaine Dulac. 1928. Illustration reproduction courtesy of Light Cone.*

in the first floor, and saw shadow plays—a glass flower world, in which all the effects worked towards the effect of a shadow; the light came once from above and once from below and then again from one side or from below. Everything colourful was dampened—nothing was harsh. Only the shadows were often harsh" (506).[129] The changing location of the lights resembles the multiperspectivism of multiple camera angles and shots as does the harsh shadows anticipate the Expressionist aesthetic. Instead of the sculptures being the focus of this composition, in a filmic manner, the shadows they cast are the main feature. Like the rhythm films of Walter Ruttmann and Dulac's *Arabesque*, the interplay of light and dark casts sculpture and light as a medium of motion, representing the teeming flow of life. Hovering between presence and absence,

129 "Da saßen wir denn ebenso wie im ersten und sahen Schattenspiele—eine Glasblumenwelt, in der alle Effekte auf die Schattenwirkung hinarbeiteten; das Licht kam mal von oben und mal von unten und dann wieder von einer Seite oder von hinten. Alles Farbige war gedämpft—nichts Grelles. Nur die Schatten waren oft grell."

FIGURE 8 *Plants from* Thèmes et Variations. *Scene still. Dir. Germaine Dulac. 1928. Illustration reproduction courtesy of Light Cone.*

the glass flowers are superseded by their shadows, becoming phantoms or representations of spirit and creative force.

While Weller's glass compositions have a great deal in common with early kaleidoscopic, experimental film, they also share some features with early nature films. Similar to the many short films on plants, Weller animates his flower compositions, speeding up the movements of plants through mechanical means: "For me, my flowers are not dead. Do you see how the precious bud opens slowly? Do you see how the stem becomes taller? Do you see how the saphire blue large leaves slowly open? A fine mechanism is within everything" (503).[130] Just as Weller uses movement to justify life in his glass plants, a short nature film called *Pflanzen leben* (*Plants Live*) from the 1920s uses time-lapse

130 "Für mich sind meine Blumen nicht tot. Siehst Du, wie sich langsam die köstlichen Kelche öffnen? Siehst Du, wie die Staubgefäße größer werden? Siehst Du, wie die saphirblauen großen Blätter langsam sich aufklappen? Eine feine Mechanik steckt da überall drin."

photography to make visible plant movement and life. Early twentieth-century nature films also framed plant movement as evidence of a plant soul. In a film from 1922, *Die Seele der Pflanzen* (*The Soul of Plants*), intertitles instruct the viewer to read plant movement as the traces of a soul: "And yet a soul lives in the body of a plant!" ("Und doch wohnt eine Seele im Körper der Pflanze!"), and "Just like a human, the chaste Mimosa quakes under the feeling of pain!" ("Gleich dem Menschen 'erbebt' auch die 'keusche Mimose' im Gefühl des Schmerzes!"). The feature length film, *Das Blumenwunder* (1926), which I discussed in more detail in chapter two, condenses the many discourses of these early nature films. The film calls on the viewer to see the accelerated movement of the plants as life; it brings to the viewer's attention the technical side of film; and of course connects movement with life.

Scheerbart's program for visual music belongs to his greater utopian vision for revolutionizing living spaces and humankind's relationship to nature through "moving architecture" ("bewegliche Architektur"). As described in *Münchhausen und Clarissa*, moving architecture has a similar effect to a "moving kaleidoscope" ("bewegende Kaleidoskope"), and the two are often brought together by providing the opportunity for, "always new, exquisite impressions" ("immer neue köstliche Eindrücke"). In "Flora Mohr," Weller's labyrinth, the jungle, is a concrete example of dynamic architecture as infinitely original space: "And all around moving mirrored-walls, that are continuously and gradually being positioned otherwise, so that one doesn't recognize the area, when one returns" (Und ringsum bewegliche Spiegelwände, die sich immer langsam und allmählich anders stellten, sodaß man die Gegend nicht wiedererkannte, wenn man zurückkam," 512). He continues to describe the effect these mirrors have on the perception of the trees: "And then, we wondered at the colossal, blue palm leaves at a small height, and we saw all around, how the other huge plants, which were mostly seen in mirrors, became time and again bigger and smaller" (514).[131] The moving mirrored-walls have the effect of integrating the glass compositions into the architectural features, allowing the plants to be read as a part of the architecture, while creating the illusion that the static glass plants are growing and shrinking. This blending of moving architecture and the glass plants belongs to Scheerbart's intentions for art to mimic the movement throughout the cosmos:

> On a star that does not stay still for a moment, we go not only together with the sun, racing rapidly onwards in a brilliant curve; we turn also

131 "Und dann bewunderten wir die kolossalen blauen Palmenblätter auf einer kleinen Anhöhe, und wir sahen ringsum, wie die anderen Riesengewächse, die zumeist in Spiegeln zu sehen waren, immer wieder bald größer und bald kleiner wurden."

> always around one another—the Earth around the sun, the moon around the earth and so on. Therefore, we must also strive to create a similar play of movement and rotation on the surface of our earth. It is really too uniform, to sit always in one place and with that to always enjoy the same view. We must bring movement in our enjoyment of nature and art. And you will have been completely convinced that it was successful at our World Fair in Melbourne, to bring movement in our whole life. (19)[132]

This architectural feature of moving walls in the interior and the exterior reappears in Scheerbart's later architecture manifesto, *Glasarchitektur*,[133] as a practical means of creating dynamic and continually new space.[134] Built in a park, the walls would create a "new architectural significance" ("neue architektonische Bedeutung"). To which Scheerbart adds, "And continually variable, this would be new" ("Und stets variabel wäre dieses Neue," 67). An expansion of architectural features to the surrounding environment would for Scheerbart result in a reconfiguration of our relationship to the natural world and to our use of architecture as an enhancement of the world's natural beauty.[135]

132 Auf einem Stern, der nicht einen Augenblick im Weltenraume stillsteht; wir fahren nicht nur mit der Sonne zusammen rasend rasch in einer großartigen Kurve weiter, wir drehen uns auch immerzu umeinander—die Erde um die Sonne, der Mond um die Erde und so weiter. Demnach müssen wir auch darauf bedacht sein, auf unsrer Erdoberfläche ein ähnliches Fahrten- und Drehungsspiel zu veranstalten. Es ist doch allzu einförmig, immerzu auf einem Punkte zu sitzen und dabei immerzu dieselbe Aussicht zu genießen. Wir müssen Bewegung in unsre Natur- und Kunstgenüsse hineinbringen. Und Sie werden sich wohl überzeugt haben, daß es auf unsrer Weltausstellung in Melbourne gelungen ist, Bewegung in unser ganzes Leben zu bringen."

133 Glasarchitektur is still read by architecture students and figures prominently within the history of architecture. Scheerbart and Bruno Taut, known for his innovative glass designs, had a great influence on one another and collaborated on small glass house for the Cologne Werkbundausstellung in 1914.

134 "Ähnliches [to the movable walls in Japanese houses] läßt sich durch verstellbare und verschiebare Glaswände in den Wohnräumen der Glasarchitektur erreichen. Bringt man die verstellbaren Glaswände [...] auch im Park an [...] eine neue architektonische Bedeutung geben. Und stets variabel wäre dieses Neue" (67).

135 Scheerbart seems to frame his moving glass architecture not in opposition to the natural but in excess of it. In another small chapter from his manifesto, he visualizes a world without brick architecture and replaced by glass architecture. As he sees it, it would be, "as if the earth dressed itself with a diamond and enamel jewelry" ("als umkleidete sich die Erde mit einem Brillanten- und Emailschmuck"). He continues: "We would have then a paradise on earth and would not need to look up with longing at the paradise in the sky ("Wir hätten dann ein Paradies auf der Erde und brauchten nicht sehnsüchtig nach dem Paradiese im Himmel auszuschauen," 42).

Furthermore, as is apparent from Weller's labyrinth, moving glass walls would also have the capacity to reveal the movement in the natural world invisible to the naked eye.

I have tried to show how Weller's glass gardens were evidence of Scheerbart's concerns with the technical aspect of creating art and how this did not preclude the spiritual. In fact, technology in many cases enabled Weller to give his glass landscapes the feeling of life through light and a mechanical apparatus that moves the flowers. The point may be that artwork is not the sum of its parts, but rather it is the force of experience and the combination of new associations that give it its soul. Even if its heart is mechanical, the glass flower is not reducible to its machinery. Instead, there is a hopeful image of technology here, one that sees technology as a means to create the ever more spiritual body, a body that wavers between the world of ideas and the material. Glass becomes a tool to transform nature, to replace our reliance on the material body with the ever better, the ever new, but the nature of such a quest leaves the perfection it seeks as incomplete—in a fragment on *Münchhausen und Clarissa*, Walter Benjamin called this impulse in Scheerbart's work the "utopia of the body" ("Utopie des Leibes," 148). The solemn sincerity of Scheerbart's vision of a better world is masked by the many narrative frames in "Flora Mohr" that reveal his cautious hopes and hide his vulnerability. But the expectation that these fantasies are realizable is apparent in Scheerbart's matter-of-fact manifesto, *Glasarchitektur* and in "Flora Mohr."

The glass flowers here are also traces of Scheerbart's struggle to push art past its slavery to verisimilitude and consequently the established art conventions. Even as the influence of different literary and art movements are present in Scheerbart's story—one can read traces of Expressionism, Symbolism, Art Nouveau, Romanticism, and science fiction—the story does not fit neatly into either one of them. Instead it reads paradoxically as self-defined, original and dependent upon art and literary history to be distinct. Likewise, the glass flowers would hardly seem so alive and pure, shimmering with light, if it were not for Flora and her "dark" objections against her uncle's play of form and light. Reality gives Weller's living landscapes of fantastical flowers a reason to exist as the very possibility of change.

CHAPTER 3

Empathetic Media: Film and the "Gestures" of Plants in *Das Blumenwunder* (1926)

> Suddenly, it has become a general need to give the inner experience an unmediated—bodily expression. Over night, it has become commonplace that language and concepts are inadequate. [...] In the face of this undoubtedly justified mistrust of words, the silent arts made popular the gesture, in which the spirit—unmediated—is being embodied without the interposition of the rational conceptual. These arts call themselves dance, pantomime and film.
>
> BALÁZS, "Tanzdichtungen", 109[136]

∴

In the previous chapter, I argued that William Weller's imaginary moving gardens in "Flora Mohr" shared their emphasis on imitating the dynamic life of nature (as opposed to its outward appearance) with other early 20th-century forms of visual culture, in particular dance and abstract film. If both dance and film often referred to plants, I argued, this was hardly by chance given the upsurge of interest in the dynamic life of plants hidden beneath their outward appearance. But plants were not simply an analogy for abstract filmmakers such as Germaine Dulac. They were also the object of an entire series of scientific and cultural films in the Wilhelmine and Weimar cinema, which featured endless images of plants and flowers moving in time-lapse. In this chapter, I want to focus on one such film that made a deep impression on both artists and scientists alike: *Das Blumenwunder* (*The Miracle of Flowers*) from 1926. Originally conceived as an advertisement for fertilizer, *Das Blumenwunder*

136 "Plötzlich ist es ein allgemeines Bedürfnis geworden, dem inneren Erlebnis unmittelbar-körperlichen Ausdruck zu geben. Über Nacht ist es zu einem Gemeinplatz geworden, daß die Sprache und die Begriffe unzulänglich seien. [...] [D]ieses zweifellos gerechtfertigte Mißtrauen unseren Worten gegenüber [hat] die stummen Künste der Gebärde, in denen der Geist unmittelbar, ohne Zwischenschaltung des rationell Begrifflichen verkörpert wird, sehr populär gemacht. Diese Künste nennen sich: Tanz, Patomime und Film."

 | DOI 10.1163/9789004327177_005

eventually became one of the most celebrated *Kulturfilme* of the 1920s, one that incorporated several well-known practitioners of the Weimar dance scene. Like Weller's spectacle in *Flora Mohr*, the film relies on movement-based media—here film and dance—to depict the life of plants, and recognizes the limits of language to express the unseen dynamism of a blossoming flower. But instead of attempting to transcend the body through the weightless transparency of coloured glass and light, *Das Blumenwunder* presents the viewer with a deliberate return to the kinetic body as the medium of a shared, lived experience with plants. In what follows, I argue that *Das Blumenwunder* employs specific ideas about embodied aesthetics at work in both film and dance theory in the 1920s in order to promote a particular type of spectatorship: namely an empathetic relation to plants—understood as dynamic, living beings—beyond the conceptual language of science.

Das Blumenwunder has not yet gathered the same attention as other Weimar film classics and has not been widely seen outside the archives, although this is gradually changing.[137] Divided into five acts through intertitles, the film is a mix of dramatic, scientific and educational elements typical of a *Kulturfilm* (culture film). Set in a garden, the film begins with young girls running and playing in between flowerbeds. When one of the girls picks a few flowers and refuses to share, their harmony quickly disintegrates into fighting and wanton destruction of the flowers. Then, a fairy appears, Flora, the protector of flowers, and interrupts their casual disregard for the plant-life in order to teach the children how to see the flowers as living beings like humans. The second act very much resembles the nature films circulating at this time in Europe, showing hands manipulating plants as well as time-lapse images of plants moving. The third and fourth acts mostly consist of time-lapse images of plants, which intermittently dissolve into flower dances performed by the Berlin State Opera Ballet and their soloists. The fifth and final act, entitled "The Song of Becoming and Passing away" ("Das Lied vom Werden und Vergehen"), begins with a flower dance and subsequent flower images, but differs from the previous two acts by repeatedly dissolving from time-lapse images of plants to the now empty garden from the first act. The available copies of the film end fairly abruptly with a cactus blooming, leaving the remaining few minutes open to speculation.

137 In cooperation with ARTE TV, the Bundesarhiv-Filmarchiv, Deutschlandradio Kultur and the cinema Babylon, the Muthesius Kunsthochschule, Kiel, screened *Das Blumenwunder* on various dates in 2011 and one in November 2013. A live orchestra, the Norddeutsche Sinfonietta, provided the accompaniment reconstructed from the original music by Eduard Künneke. A dossier with extra film materials including reviews is also to-date available on the ARTE TV website.

Das Blumenwunder's première on February 26th, 1926, at the Piccadilly Cinema in Berlin was warmly greeted by the press with mostly positive reviews. The *Berliner Morgenpost* named the film a "great marvel" ("großes Wunder") and "perhaps the sensation of the year" ("vielleicht die Sensation dieses Jahres," press clippings, Archiv der Deutschen Kinemathek). According to a review from the *Filmkurier*, the nature images became through this film an "artistic event" ("künsterlischen Erlebnis," press clippings, Archiv der Deutschen Kinemathek). In the same review, it was estimated that over 70,000 people saw this film in 63 sold out screenings. There were, to be sure, some critiques. One reviewer questioned the pedagogic value of the third to fifth acts, noting that little information was provided through the images and the dancers, while other reviewers complained about the seemingly endless number of time-lapse images or the "ugly" expressionist dances (*Lehrfilme* 22).[138] On the whole, however, the film continued to exert a broad appeal going well beyond the world of educational film. Representative, in this sense, are the remarks of Rudolf Arnheim, who would describe *Blumenwunder* in his book *Film als Kunst* (*Film as Art*, 1932) as "definitely the most exciting, fantastical and beautiful film... that was ever filmed" ("sicherlich der aufregendste, phantastischste und schönste Film... der je gedreht wurde," press clippings, Archiv der Deutschen Kinemathek).

Reading such critiques today, what stands out above all is a repeated desire to see in *Das Blumenwunder* a model for another relationship between the viewer and the filmic spectacle. For example, *Der Montag Morgen* wrote: "One *lives* das Blumenwunder, *feels* the heart beating with the rhythm of the unfolding, and *senses* that these plant forms belong with animals and humans—they belong to the great realm of *ensouled, suffering*, and *active* beings in nature." ("Man *erlebt* das Blumenwunder, *fühlt* das Herz im Takt mit dem Rhythmus der Entfaltungen schlagen und *empfindet* die Zugehörigkeit dieser Pflanzengestalten zu Tier und Mensch—zum großen Reich *beseelter leidender* und *handelnder* Naturwesen," 10, italics mine).[139] This comment

138 One review from Lehrfilme noted negatively the unending number of time-lapse images of plants or the seemingly ugly style of the expressionistic dancers. The same reviewer critiqued the pedagogic value of the third to fifth acts, noting that little information was provided through the images and the dancers, although he did find the framing narrative instructive, teaching children not to mistreat plants (22). See Dr. M-1, "Kultur- oder Lehrfilm? Kritische Betrachtungen zum 'Blumenwunder'." (*Der Lehrfilm*: Beilege zu „Der Filmspiegel." *Kinematographische Monatsheft*, Berlin, July 1926. *Archiv der deutschen Kinemathek*).

139 *Tägliche Rundschau*: "Wir erleben den Pulsschlag der Pflanze, bewundern andächtig die Allmacht der Natur..." ("We experience the beating of the pulse of the plant, marvel

and others like it have caught upon the unique viewing experience that the time-lapse images of plants present. While at times the reviewers focus on the revelatory power of film to depict plant movement, they also suggest that such depictions could call forth reactions of empathy in viewers. For Rudolf Arnheim, this process relied upon the accelerated pace of time-lapse images to reveal that plants could display "gestures" ("Mimik"), and "expressive movements" ("Ausdrucksbewegungen") analogous to those of humans and animal.[140] Focussing on *Das Blumenwunder*, I wish to consider how the time-based media of film and dance could be employed to kinetically teach viewers to adopt an empathetic relation to plants as living beings.

Before *Das Blumenwunder* reaches this bodily mode of communication and of viewing film, it prepares the way explicitly and implicitly for viewers to understand plant movements and film without articulated conceptual language. The film program for the première states outright that language is inadequate to describe plant movement: "One cannot describe these movements, this searching, struggling, and grasping of the creeper. [...] The written language is missing the words. The moving image illustrates for us like a holy book" (Film program np).[141] Even though language is inadequate to convey the meaning of the plant movements, the first two acts still rely on a cognitive understanding of the images through intertitles to attribute meaning. Conceptual language is combined with familiar film forms of the *Kulturfilm*, the nature film and even references to early attraction cinema to teach the viewers how to understand and see the last three acts without the aid of intertitles. In this progression from a communication model based on conceptual language to an embodied language, the assumptions of culture and nature

---

with rapt attention at the almighty power of nature"); Vorwärts: "... durch diesen Film bewußt werden, wie die Pflanze wächst und den Umkreis ihres Lebens durchschreitet ..." ("... being made aware through this Film how the plant grows and strides through the circumcircle of its life ..."); *Deutsche Allgemeine Zeitung*: "Es ist erschütternd zu sehen, wie die Pflanzen streben, gehemmt werden, leiden, verzweifelt sind ..." ("It is shocking to see how the plant is striving, hemmed in, and plagued with doubts") (press clippings, *Archiv der Deutschen Kinemathek*).

140 The connection between film techniques becomes obvious through Arnheim's complete statement: "... bei diesen Aufnahmen hat sich herausgestellt, daß die Pflanzen eine Mimik haben, die wir nicht sehen, weil sie mit zu langsamen Zeit rechnet, die aber sichbar wird, wenn man Zeitrafferaufnahmen verwendet. ... die Pflanzen ware plötzlich lebendig und zeigten Ausdrucksbewegungen von genau denselben Art, wie man von Menschen und Tieren kennt."

141 "Man kann diese Bewegungen nicht beschreiben, dieses Suchen, Kämpfen und Greifen der Kletterpflanzen. [...] Es fehlen der geschreibenen Sprache die Worte. Das schildert uns wie ein heiliges Buch das bewegte Bild."

film behind the strict categorization of people as subjects and plants as objects are broken down and reassigned. This allows for the opportunity of another subjectivity, not measured by human standards of intelligence, but rather from a plant-based perspective of movement.

The keys to interpreting this plant perspective lie in two movement-based media, dance and film, which can provide meaning without resorting to an articulated language when attempting to interpret or "translate" the subjectivity of plants into the world of human perception. Dance becomes a model of kinetic learning that teaches the first step to a different form of understanding, one that can only come about through a mimetic connection to natural movement. Through imitation, dance transposes the movements of plants to the human body, so that the movement can be felt by the viewers in addition to being observed. The plant dances become a mode of communication between the voiceless natural world and the viewer's non-cognitive experiences of the world. Film not only acts as the means to showcase this mimetic model, but also enhances the similarities in movement between the dancers and the plants through time-lapse photography. Gradually, the film eases from a cognitive and visual interaction with the images into a bodily perception, encouraging the spectators to engage with the flowers through their bodies, and therefore, empathize with the movement and life of plants.

As acknowledged in the first two acts, conceptual language has great power to explain ideas, disseminate knowledge and tell stories. Yet, language also has played a large role in the disembodiment and suppression of other sensory-based ways of knowing in Western culture. Some scholars have traced this suppression of sensuous knowledge back to the creation of the Phonetic system of signs and its adoption by Plato, Aristotle and other Greek thinkers.[142] According to Kevin Rathunde, it was Plato's universals and their disconnect from the natural world that helped to form thought in the West and led eventually to Descarte's mind-body dualism (191). He goes on to conclude that this separation led to a devaluation of the body and the view that sensory information is a potential source of illusion.

Without delving too deeply into discussions on language and the body, it is still important to note that the relationship of spoken language to the body is closer than that of the written word to the body.[143] While the speech act can

142 See Kevin Rathunde "Montessori and Embodied Education," 190; David Abrams *The Spell of the Sensuous: Perception and Language in a More-than-Human World*, 106.

143 Tony Jackson presents in *Writing and the Disembodiment of Language* a nuanced view of language as rooted in biology and also culturally determined. While not fully determined by reality, language is also not completely arbitrary nor relativistic as forwarded by many post-structuralists. Instead, writing as a technology and model has caused speech

also assume some aspects of the written word as an object, the act of speaking remains rooted in the body as that which must be experienced to exist. For this reason, the inclusion of an image of a man speaking in the fourth act of *Das Blumenwunder* is significant. The lack of accompanying sound—his speech—emphasizes the role of the body as meaningful during the act of speaking. Additionally, the body language pictured here provides a key for reading the other images of the dancers and the accelerated images of the growing plants as a form of communication. The small, quick gestures of the man speaking (Figure 9) are echoed in the subsequent image by the twitchy movements of flowers growing. Analogous to the man's gestures, the movement of the flowers comes to be seen as body language.

The failure of language to convey the excess meaning of the lived experience or body language were of concern to writers and theorists from the turn of the century well into Weimar Republic. The symbolist writer and thinker, Maurice Maeterlinck, described the failure of language to express experience as dynamic and living. For film theorist, Béla Balász, the emphasis on the written word had overshadowed other means of communication and divorced the soul from its body. Part of film's promise, according to Balázs, lay in its perceived potential to restore communication at the level of the body and its gestures, and thereby to enrich an impoverished humanity—one characterized by a dematerialized and abstracted relationship to the natural world. The desire was not to replace conceptual language as a means of communication but to displace its hegemony over expressions of the human self.

One finds this desire thematized frequently in early 20th-century German culture. Béla Balázs speaks of communication through gestures as that which has been lost. Similarly, Franz Kafka's short story, "Ein Bericht für eine Akademie" ("A Report for an Academy," 1920), thematizes this lost access to the natural world through an ape that acquires the ability to speak first through imitating the humans in his environment.[144] Both Balázs' theory of gestures

to seem at times inadequate, just as the speech act composes a complex communication act involving not only bodies, but social context and orality itself. He concludes that written language and its status as an object has come to subsume all forms of language and created objects out of them.

144 In his report, the ape can remember a time before he had access to language: "Ich kann natürlich das damals affenmäßig Gefühlte heute nur mit Menschenworten nachzeichnen und verzeichne es infolgedessen, aber wenn ich auch die alte Affenwahrheit nicht mehr erreichen kann, wenigstens in der Richtung meiner Schilderung liegt sie, daran ist kein Zweifel" ("Naturally, I can nowadays only retrace the former ape-like feeling with human words and draw it as that which follows. But if I can no longer access the old ape-truth, at least my illustration gives the right impression. Of that there is no doubt" (206). As in the

FIGURE 9 *A Man Speaking from* Das Blumenwunder. *Scene still.* BASF AG *Unterrichtsfilm, Verlag wissenschaftlicher Filme, 122–1926. Illustration reproduction courtesy of* Filmarchiv-Bundesarchiv.

and Kafka's pseudo-Darwinian tale imply an earlier, more physical way relating to the world and to others before conceptual language took hold. But even more pertinent for understanding the thematization of communication and the visual strategies of *Das Blumenwunder* is the work of Walter Benjamin, who understood the mimetic faculty—i.e. our ability to identify similarities—as the remaining trace of a once great power to live in the world mimetically (Benjamin, "Doctrine of the Similar," 69).[145] Where others saw interacting with the world mimetically as an earlier stage of language, Benjamin believed that we were once able to read and understand the natural world through symbols as we now read language. The subsequent philosophical career of Benjamin's concept in the writings of Adorno and the Frankfurt School—

film, imitation also serves the function to learn how to communicate with the sailors on the ship as they brought him to Europe from Africa.

145 Benjamin writes: "The gift which we possess of seeing similarity is nothing but a weak rudiment of the formerly powerful compulsion to become similar and also to behave mimetically" (69).

where mimesis held out the promise of an alternative to the instrumentalizing force of conceptual thought and language—is well known.[146] But the concept can also help us to approach a prevalent current of Weimar visual culture in photography and film: namely the preoccupation with visual analogies. A good example can be seen in a journal such as *Der Querschnitt* (1921–1936), where the editors never tired of highlighting the visual similarities between disparate items—animals, machines, people, artworks, etc.—in photomontages that forewent all textual explanation.[147] It should perhaps come as no surprise that plants often figure within these constellations. One montage from a 1925 edition of *Der Querschnitt* highlights the similarities between a dancer and a group of orchids (Figures 10 & 11), while another juxtaposes the mouth of a roaring tiger with a photo of an orchid.

Such visual analogies would also form a recurrent motif of 1920s cinema in films such as Hans Richter's *Zweigroschenzauber* (1929), where a magician employs a series of dissolves to highlight the surprising similarities between conceptually distinct objects (e.g. a bald head and a full moon) before the enraptured gaze of his audience. Like Richter's film and the other examples of visual analogies, *Das Blumenwunder* employs the same strategies to circumvent a rationalized and objectifying view of nature through a model that attempts to reconcile conceptually disparate objects.

While the contrast between conceptual and embodied language as it plays out in this film is the focus of this chapter, the difference between the natural rhythm of plants and the "*Takt*" of measured time is also a dominant theme in this film and deserves some attention. *Takt* refers to the artificial and mechanical time in clocks, pendulums and other mechanical devices with staccato movements (Blankenship 98). *Takt* also has a secondary meaning, referring to a group of dancers keeping in time together (99). In the first and—to the date

146 Adorno adopts Benjamin's concept of mimesis and seems to suggest that it is a solution to the rationalized and instrumentalized use of language as a tool of power without clearly defining what he means by mimesis. Ernesto Verdeja interprets Adorno's concept of mimesis as a problematic reconciliation with nature and of subjectivity and objectivity. Particularly useful for interpreting mimesis as it appears in Das Blumenwunder is both mimesis as symbolic understanding of movement in nature yet physical and as counteracting a rationalized and objectifying view of nature through a model that attempts to reconcile the two.

147 On the use of photomontage in *Der Querschnitt*, see especially Kai Sicks, "Der Querschnitt: oder die Kunst des Sportreibens," in Leibhaftige Moderne, ed. Michael Cowan and Kai Sicks (Bielefeld: Transcript, 2005), 33–47; Michael Cowan, "Cutting through the Archive: Querschnitt Montage and Images of the World in Weimar Visual Culture," *New German Critique* 120, vol. 40, no. 3 (2013), 16–21.

FIGURE 10 *Orchideenblüten, photograph from Albert Renger-Patzsch,* Der Querschnitt. *6 March 1926: 41.* Illustrierte Magazine. *Web. 12 Aug 2013.*

FIGURE 11 *Tanz im Freien, photograph from Gerhard Riebicke,* Der Querschnitt. *October 1925:* 57. Illustrierte Magazine. *Web. 10 Jan 2014. Illustration reproduction courtesy of Gerhard Riebicke / Bodo Niemann Berlin.*

of writing—the only journal article published on *Das Blumenwunder*, Janelle Blankenship contextualizes the film in its contemporary art and science discourses with rhythm as a connecting thread.[148] She argues that this film offers the possibility of insight into the natural rhythm of the plant world through time-lapse photography and the contrasting *Takt*. She draws on the work of 19th century biologist and semiotician, Jakob von Uexküll, who, when arguing against the machine model of the natural world, extolled the possibilities of film to reveal the subjective time of animals and plants by stretching or shortening the human minute. These same possibilities of film also appealed to numerous avant-garde artists such as the filmmaker Germaine Dulac, who delighted in and felt inspired by film's ability to reveal the rhythms of the plant and animal world.[149] Blankenship also touches on the possibility of an interactive film spectatorship and the dancers' mimicry of the blossoming flowers. It is from these two points that I intend to expand her interpretation of the film.

Jakob von Uexküll's biosemiotics suggest a way to view plants not just as objects but also as potential subjects.[150] He uses music metaphors, composition and counterpoint, to interpret how an animal interacts with its surrounding environment. Every possible relationship is determined by a meaning carrier, a scent in one of von Uexküll's examples, and a meaning receiver, a moth in the same example. Their relationship, according to von Uexküll, is a duet and a composition, which is determined by a meaning rule (100). The scent of a female moth holds meaning for the male moth and directs his attraction. The meaning carrier then forms the shape of the meaning receiver or its counterpoint. Von Uexküll's main argument seeks to escape an anthropocentric view of nature in favour of one where everything has a subjective viewpoint from bacteria to flowers. His line of thought provides a method to question the meaning of every interaction from the perspective of the other.

---

148 At a screening of the film, Ines Lindner moderated a roundtable with the culture studies scholar, Peter Berz (Humboldt University), the film scholar Gertrud Koch (Freie Universität Berlin) and the Künneke expert Sabine Müller (Cologne).

149 Dulac's concept of visual music is discussed in the previous chapter on Scheerbart's novella, "Flora Mohr." Her vision of a film aesthetic that was composed of rhythmic visuals without conceptual language is also pertinent for *Das Blumenwunder* and is discussed further in Blankenship's article.

150 Von Uexküll's treatise *Umwelt und Innenwelt der Tiere* (*The Subjective Universe of Animals*, 1909) argues that the surrounding environment influences the individual animal and vice versa. His treatise forms the cornerstone of contemporary scholarship on ecocriticism and literature. See Kay Milton, *Loving Nature: Towards an Ecology of Emotion* (London: Routledge, 2002, Ebrary, Web, 9 April 2014).

For my purposes here, it hints at a plant-based way to interpret the interaction of humans and plants in *Das Blumenwunder*.

German aestheticians coined the term *empathy* at the end of the 19th century to describe acts of viewing characterized by a physical connection between the artwork and the viewer.[151] In his study, *Abstraktion und Einfühlung* (Abstraction and Empathy), Wilhelm Worringer described the modern aesthetic as having taken a step from aesthetic objectivity to aesthetic subjectivity, which meant that any examination of art needed to start with the "behaviour of the observing subject" ("Verhalten des betrachtenden Subjekts") instead of the "form of the aesthetic object" ("Form des ästhestischen Objektes," 2). Such a method results in a theory, which Worringer names "lessons in empathy" ("Einfühlungslehre"). He defines this aesthetic experience as follows: "Aesthetic pleasure is objectified self-pleasure. Aesthetic pleasuring means for me to take pleasure in myself in a sensual object that is differentiated from me—to feel myself in it" (4).[152] For Worringer, viewing art involves a dynamic subject whose ability to transpose the self into the physical object and to feel it comprised the physical experience. When applied to viewing both film and dance, the experience transforms a purely passive or disembodied one to one in which the images result in physical sensations.[153]

Such notions of embodied aesthetic experience worked out in modern theories of empathy have continued to play an important role particularly in dance studies. Dance studies scholars, John Martin and Susan Foster, for example, have seen the usefulness of this early connection of empathy with physical experience as a way of accounting for the unique physicality of viewing

151 Susan Foster provides a short history of the word empathy in her introduction to *Choreographing Empathy*.

152 "Aesthetischer Genuss ist objektivierter Selbstgenuss. Aesthetisch geniessen heisst mich selbst in einem von mir verschiedenen sinnlichen Gegenstand geniessen, mich in ihn einzufühlen".

153 Friedrich Nietzsche also wrote on empathy in *Daybreak* (1881): "Empathy—To understand another person, that is, to imitate his feelings in ourselves, we do indeed often go back to the reason for his feeling thus or thus and ask for example: why is he troubled?—so as then for the same reason to become troubled ourselves; but it is much more usual to omit to do this and instead to produce the feeling in ourselves after the effects it exerts and displays on the other person by imitating with our own body the expression of his eyes, his voice, his walk, his bearing (or even their reflection in word, picture, music). Then a similar feeling arises in us in consequence of an ancient association between movement and sensation, which has been trained to move backwards or forewards in either direction. We have brought our skill in understanding the feelings of others to a high state of perfection and in the presence of another person we are always almost involuntarily practising this skill" (89).

dance.[154] In a series of four lectures, Martin spoke of the contagion of seeing physical movement as metakinesis;[155] through kinetic sympathy or more concretely described as muscular sympathy, the audience responds to the dance by feeling in their musculature the dancer's movements. Susan Foster broadens Martin's notion of kinetic sympathy to include the communication not only of the dancer's emotional and aesthetic experience but also of culturally defined modes of embodiment such as gender and ethnicity.

The latter observation also has relevance for a film like *Das Blumenwunder*. For the empathetic spectatorship the film calls forth can also be understood in gendered terms as a mode of feminine spectatorship—one that seeks to replace the masculinized order of knowing nature with an embodied relation. Film theory has long been concerned with the ways in which gender positions are played out through the filmic gaze. In particular, following Laura Mulvey's famous model of the sadistic and disembodied male gaze in classical Hollywood cinema, several theorists in the 1980s sought to articulate models of viewing coded as "female." Gaylyn Studlar, for example, posited the existence of a "masochistic" spectator, who becomes a child in front of the dream screen and returns to a sense of unity with the mother.[156]

More recently, film theorists such as Laura Marks and Vivian Sobchack, although not working explicitly within a gender model, have sought to rethink filmic reception in general as an embodied affair in opposition to the disembodied male gaze of theories such as Mulvey's. Their project is an effort

---

154 Other scholars include Walter Sorell who described kinesthetic sympathy as "the inexpressible dialogue occurring between dancers and between dancers and the audience; the experienced sensation over and above what can be reiterated in words," quoted in *Dancefilm* (12). Mary M. Smyth leans toward a scientific approach to kinesthetic sympathy in her article "Kineasthetic Communication in Dance" (19). Carrie Lambert gives a brief historical summary in "On Being Moved: Rainer and the Aesthetics of Empathy" (45–6).

155 John Martin defines metakinesis in *The Modern Dance* as: "Movement, then, in and of itself is a medium for the transference of an aesthetic and emotional concept from the consciousness of one individual to that of another" (13). In a 1946 review, he describes the experience of muscular sympathy as feeling sympathy in one's muscles the effort seen in another's (22).

156 Studlar sees the screen as transforming the spectator into child and the screen into the nurturing mother: "In a sense, these same wishes are duplicated by the film spectator who becomes a child again in response to the dream screen of cinema. This dream screen affords spectatorial pleasure in recreating the first fetish—the mother as nurturing environment" (787). In the real and mirror stages, the child has no control over the image of his ego and is analogous to the relationship of the spectator to the screen and images: "The object/screen/images cannot be physically possessed or controlled by the spectator. The spectator's 'misapprehension' of control over cinematic images is less a misapprehension than it is a disavowal of the loss of ego autonomy over image formation" (788).

to recognize that films engage not only the senses of sight and hearing but also the sense of touch—they are felt. “Haptic visuality,” as Marks calls embodied viewing, “draws from other forms of sense experience, primarily touch and kinaesthetics,” causing the “viewer’s body [to be] more obviously involved in the process of seeing than is the case with optical visuality” (332). Optical visuality, in contrast, promotes distance between the image and the viewing, allowing space for the viewer to identify with the image and project herself onto the object (335–6). Haptic visuality shifts the relationship of the viewer to the image away from the divided subject-object to a merged subjectivity—what Sobchack calls the cinesthetic subject. In contrast with identifying with someone or something on the screen, the cinesthetic subject experiences film through the sensorium and “without a thought.” According to Sobchack: “We, ourselves, are subjective matter: our lived bodies sensually relate to ‘things’ that ‘matter’ on the screen and find them sensible in a primary, prepersonal, and global way that grounds those later secondary identifications that are more discrete and localized” (65). This bodily experience of film viewing commingles with the cinematic representation to the extent that meaning arises from a conjunction rather than either (67). The two forms of viewing, haptic and optical, are not mutually exclusive—often shifting frequently and swiftly throughout a film.

In *Das Blumenwunder*, the transition from optical visuality to embodied empathy is enacted by the very progression of the film’s acts. During the first act, the film depicts quite distinctly a rationalized relationship to space, movement and in language. As the film opens, there is a definite sense of order and sectioning of the garden space and the girls’ relationship to it. This is reflected in the clear separation between the flowerbeds and the paths in between, which in turn affects a separation between people and nature. As becomes clear, in one shot taken from a high camera angle, the girls’ movement respects the ordered and rationalized space of the garden. They run in lines between the flowerbeds with the intent of catching the first girl, who first taunted the others into a chase. When the girl being chased trips and falls into a flowerbed, she stands up with two bunches of flowers in her hands. Both she and the girls surrounding her express wonder at the flower’s beauty, followed by a desire to possess this beauty. There ensues a fight to obtain the flowers that recalls a long history of viewing plants as objects, one embodied by the classification system of Linneaus. By invoking a familiar relationship to nature and to space, these introductory images act as a persuasive device to prepare the audience to see similarities between plant movement and dancing rather than immediately launching into a series of unexplained images of dancers and plants, which might or might not have the desired effect.

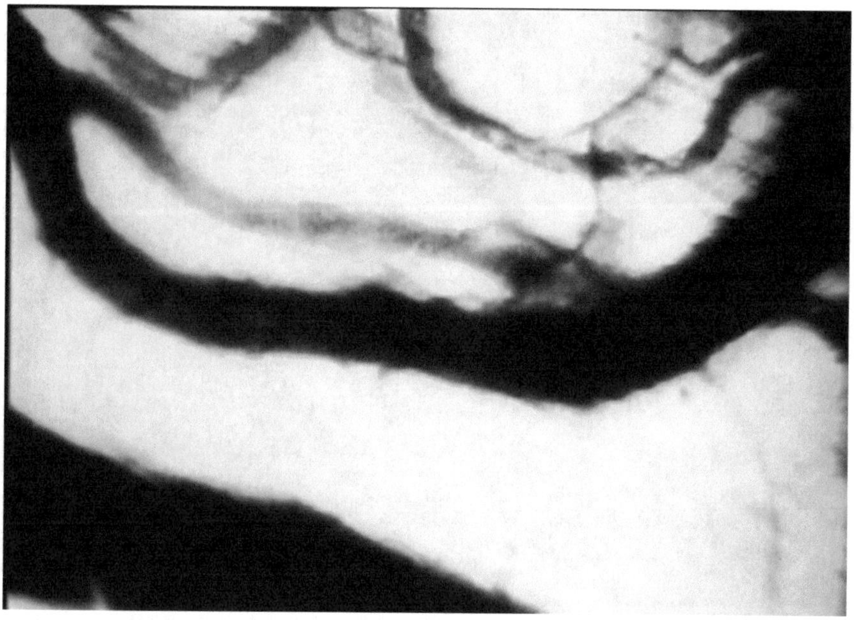

FIGURE 12 *Pulsing Veins from* Das Blumenwunder. *Scene still.* BASF AG, *Unterrichtsfilm, Verlag wissenschaftlicher Filme, 1922–26. Illustration reproduction courtesy of* Filmarchiv-Bundesarchiv.

The use of optic visuality to induce the viewer to identify with the girls is manipulated by the film to shift the viewer's relationship to the image. The short narrative in the garden is followed by a short reversal of perspective. At first, the viewer seems to be positioned as an observer of the garden and the girls through the ordering of space and the position of the camera. But just before Flora's appearance in the garden, the camera drops to the level of the girls' viewpoint and the viewer now enters into the garden and begins to share their perspective. This shift in perspective also coincides with the girls' own transformation through Flora. Her role is to teach them to perceive aspects of nature previously beyond the scope of their acquired vision in both a spatial and temporal sense. She uses a conceptual analogy between the rhythm of the pulse, clocks and finally a clip of accelerated plant growth to illustrate her assertion that: "The flowers are alive like you" ("Die Blumen haben Leben gleich euch"). The microscopic clips of pulsing veins (Figure 12) and racing blood cells (Figure 13) highlight the ability of the microscope and the film camera to translate worlds beyond the human sensorium. Introduced by zooming into a blurred image of hands, the image of the veins and blood also has a secondary effect of shifting from optical visuality to haptic visuality. The texture

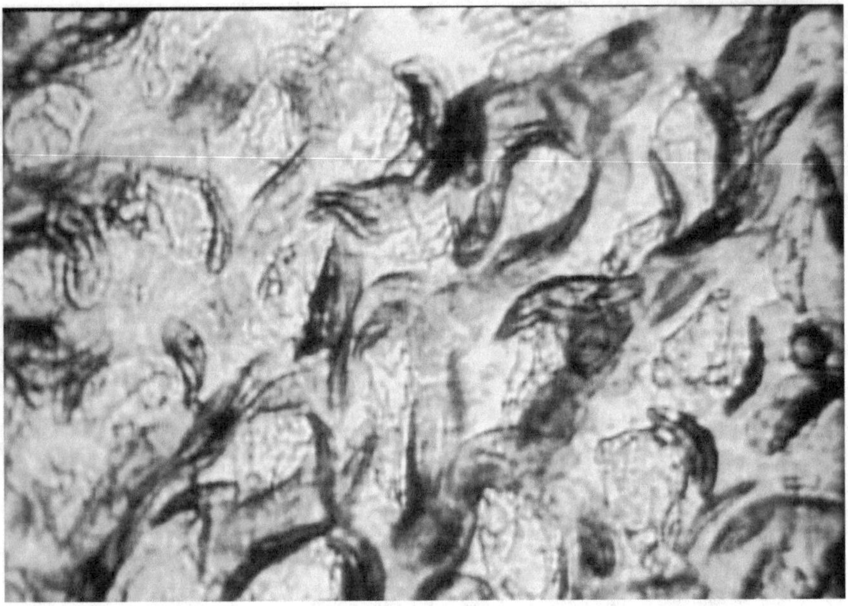

FIGURE 13 *Racing Blood Cells from* Das Blumenwunder. *Scene still.* *BASF AG, Unterrichtsfilm, Verlag wissenschaftlicher Filme, 1922–26. Illustration reproduction courtesy of* Filmarchiv-Bundesarchiv.

of the images encourages the activation of a sense of touch while the pulse of the vein encourages awareness of the rhythm of the viewer's pulse. The biological felt rhythm is then translated back to a conceptual explanation of the capacity of film to speed up time and visualized through the racing hands of a clock (Figure 14). A clock with days of the week instead of numbers and flowers on the face (Figure 15) is then used to complete the visual translation between the pace of the human pulse, the *Takt* of a clock and the slow time of plants. The subsequent clip of the accelerated movement of a flower growing is then to be understood as a combination of measured, rationalized time and biological, felt rhythm. Furthermore, the movements of the flower are meant to be read as analogous to the human body, setting the stage for later visual analogies between the dancers and the flowers.

In generic terms, Flora's role is not unlike that of the showman or lecturer from the early cinema of attractions (Figure 16)—a role still well known in the science film genre—who sought to heighten curiosity and build expectation through speech.[157] Like the cinema lecturer, Flora uses language to focus the

157 Tom Gunning discusses the role of the showman in early cinema through James Stuart Blackton, who creates in his introduction of the train film a "sharpening of expectation"

FIGURE 14 *Racing Clock Hands from* Das Blumenwunder. *Scene still.* BASF AG, *Unterrichtsfilm, Verlag wissenschaftlicher Filme, 1922–26. Illustration reproduction courtesy of* Filmarchiv-Bundesarchiv.

children's attention and subsequently that of viewers on particular connections they should notice in the images. Before the microscopic image of blood flowing through veins, she explains how to see the image: "For humans, the rhythm of life is the beating of the pulse, is the meeting of the blood cells" ("Der Lebensrhythmus des Menschen ist der Pulsschlag, ist das Tagen der Blutkörperchen"). Similarly, between the images of the clocks and the one time-lapse image of a flower blooming in the first act, she explicitly lays out the analogy between the pace of human life and plant life: "Twenty four hours, a day in the life of a person, is the second in the life of the flower" ("Vierundzwanzig Stunden, ein Tag im Leben des Menschen, ist die Sekunde im Leben der Blume"). She then intensifies curiosity through delay by announcing: first "And now look there! Enchanted by the grasp of time, your eyes will see" ("Und nun schaut her! Im Zeitgriff verzaubernd werden Eure Augen sehen"), and then

---

and "cathecting of curiosity through delay" ("An Aesthetic of Astonishment: Early Film and the [In]Credulous Spectator" 46).

FIGURE 15 *Clock with Days of the Week from* Das Blumenwunder. *Scene still.* *BASF AG, Unterrichtsfilm, Verlag wissenschaftlicher Filme, 1922–26. Illustration reproduction courtesy of* Filmarchiv-Bundesarchiv.

"The miracle of flowers should bloom in front of you" ("Das Blumenwunder soll vor Euch erblühen"). As the exhibitor, Flora thus draws back the metaphorical curtains to reveal time-lapse films of plants as objects in a display case. It is now impossible to see an unmediated image of the plants, free from her authoritative influence.[158] The reliance on conceptual language in the intertitles to explain the visual analogies demonstrates the primacy of Béla Balázs' reflections on the impact of print on human interaction. He cites the written word as the "principle bridge joining human beings to one another," and identifies a disconnect between the body and the soul: "In the word, the soul

158 As the mediator, Flora also refutes the common role of women in film as objects of display and the gaze. She is not as Laura Mulvey describes on display and nor is she a passive object of the male gaze and subject to him (840). She engages in a dialogue with the girls, during which she directs their gaze away from her as the attraction to the films: "Und nun schaut her" (And now look there). With the subsequent insertion of a time-lapse flower blooming, the girls' gaze is conflated with the film viewer's and shifts to being receptive.

FIGURE 16 *Flora Pointing from* Das Blumenwunder. *Scene still.* BASF AG, *Unterrichtsfilm, Verlag wissenschaftlicher Filme, 1922–26. Illustration reproduction courtesy of* Filmarchiv-Bundesarchiv.

has been collected and crystallized. Without the soul, the body is empty" (*Der Sichtbare Mensch*, 47).[159] The film *Das Blumenwunder* is based on the assumption that viewers, having lost the ability to use a bodily language to connect with one another, need conceptual language as an initial bridge in order to find a way back to a sensuous form of communication.

It is Flora who directs this transition between conceptual and embodied language. Just as she uses media (microscopy and time-lapse photography) to make manifest a world beyond the reach of the human sensorium, so she herself *becomes a mediator*: a "go-between" bridging the worlds of conceptual and bodily language and—ultimately—of human beings and nature. Like a spiritualist medium, Flora allows the life of plants to "appear" through the body and speak through her voice, thus attesting to their subjectivity for the humans in the film and in the cinema. Indeed, her very name, Flora, as well as her costume, decorated in flowers, identify her as a hybrid of human and

159 "Doch seit der Buchdruckerei ist das Wort zur Hauptbrücke zwischen Mensch und Mensch geworden. In das Wort hat sich die Seele gesammelt und kristallisiert. Der Leib aber ist ihrer bloß geworden: ohne Seele und leer."

FIGURE 17 *Das Blumenwunder Front Cover from Souvenir Program. 1926*. Archiv der Deutschen Kinemathek. Art.tv. *Web. 18 Nov 2013. Illustration reproduction courtesy of the* Archiv der Deutschen Kinemathek.

plant. Likewise, the cover drawing from the film program also depicts Flora as a hybrid figure, a human form that is practically indistinguishable from the flowers surrounding her (Figure 17). Her hybrid appearance indicates access to both plant and human languages, necessary to overcome their communication impasse. In her role as "go-between," Flora thus embodies the tension between plants as objects and plants as subjects at the centre of the film's project.

Even as Flora stands between the viewer and the images, she also plays the *intermediator*, an advocate, who pleads the case of plants, as it were, to the viewing human community. In her role as "protector of flowers," Flora translates the bodily language of plants to the conceptual language of humans, thereby lending to the plants another subjectivity, one capable of being registered and understood by people. The waving of a leaf or a blooming of a flower becomes their "Suffering and Struggle" ("Leiden und Kämpfen"), and more explicitly she states that plants "feel ... like you in blooming and wilting" ("empfinden ... gleich euch im Blühen und verwelken"). She interprets the language of the plants as signs of feelings both corporeal and psychological. Just as these words shape the viewer's understanding of the latter plant images, they also prepare the viewer to step out of conceptual language to understand the waving leaf from contextual clues, when the flowers no longer need rely upon Flora as their voice.

In her role as the intermediator, Flora cannot be separated from reflections on the medium of film and the film's subject matter. In this sense, she differs from other narrators in early nature films. For example, the male narrator in a similar film from 1931 entitled *Geheimnisse im Pflanzenleben* (*Secrets in the Life of Plants*) remains apart from the natural world and privileges conceptual language over images for learning about nature. Like Flora, that narrator describes plants as a "Living being ... like humans and animals—with sensations, emotions and certain senses" ("Lebewesen ... wie Mensch und Tier—mit Empfindungen, Regungen und bestimmten Sinnen," *Geheimnisse der Pflanzenleben*, censorship card, Filmarchiv-Bundesarchiv). Instead of emphasizing the visual capacity of film to teach about plants, the time-lapse images are illustrations and an advertisement for the character's forthcoming book. It is from the book, and not the film, that "everyone will learn out of it" ("jedermann wird daraus lernen," np). The time-lapse images are treated as a strictly realistic depiction of the plants from the greenhouse around them in an intertitle near the end: "Miraculous, this life of plants—how dead in comparison is every image!" ("Wunderbar, dieses Leben der Pflanzen.—Wie tot dagegen jedes Bild!"). Although he also explains the time it takes to photograph plant movement to create the time-lapse images, he ignores the mediated quality of the images, conflating the accelerated images of plants growing with plants in the natural world. In contrast to his position as apart from the natural world and from film, Flora is both embedded within the diegetic world and gesturing to films embedded further inside from her position as mediator of the image and the voice of the flowers.

Flora's role as embodied nature, as the voice of the flowers and the mediator of the image connects her into a tradition of conflating women with nature yet reframes the meaning of the association. While traditionally women have been conflated with plants to emphasize their passivity and inferior intellectual capacities, Flora's active role represents a vision of plants that no longer fits into the traditional femininity. As a dynamic visualization of plant movement, she reflects how the plants, who can search and react, also no longer fit into the traditional view of plants as passive. The interweaving of femininity and plants in *Das Blumenwunder* participates in affirming traditional femininity as closer to the natural world while opening it to new and active roles as a public advocate. While physically, Flora does not resemble the "new woman" ("neue Frau") in the Weimar republic, she is participating in a similar reframing of femininity as active and public. The version of femininity represented by Flora illustrates a far more complex engagement than the simple masculinizing of women to achieve emancipation.

Following upon the narrative frame of the opening act, the second act of *Das Blumenwunder* presents transitions into another familiar aesthetic mode: namely that of the *Naturfilm* (nature film), which uncannily echoes the girls' violence towards plants at the beginning of the film in language and human interference. Here, time-lapse images of plants filmed against a dark background are explained through intertitles that range from providing information such as the plant names (*Bananenblatt, Sichelfarn…*) to describing the images in scientific or dramatic language. Common to this section are intertitles such as the following: "But the person intends to impose the left rotation on her [the plant] with raw force" ("Aber der Mensch will ihr [der Pflanze] mit roher Gewalt die Linksdrehung aufzwingen"), to which the plant responds and "remains the victor" ("bleibt Siegerin"). Such overt references to human violence and scientific manipulation are uncommon in most *Naturfilme* of this period. Instead, the intertitles at times simply name the plants or offer dry descriptions of the images, as in the following intertitle from *Hormonwirkungen bei höheren Pflanzen*: "Effect of one-sided application of Heteroauxin paste (0.005% H.A.) on the intact oat coleoptile time speeded up 480 times" ("Wirkung einseitig aufgetragener Heteroauxinpaste (0,005% H.A.) auf die intakte Haferkoleoptie Zeitraffung: 480-fach"). In other 1920s nature films, there are often pedantic and even dramatic descriptions of the images, as in the following intertitle from *Wunder der Natur: Aus den Wurzeln kommt der Kraft* (*Miracles of Nature: Out of the Roots Comes the Power*): "So a root tip is an exceptional artistic image. The whole fate of the plant hangs off of it."[160] This film echoes Charles Darwin's fascination with the sensitivity of root tips, and epitomized in the final sentence of *The Power of Movement in Plants*, which described root tips as the plant's brain. Lastly, plant life itself was often sensationalized with violent imagery,

160 "So eine Wurzelspitze ist eine ausserst kunstvolle Gebilde. Von ihr hängt der ganze Schicksal der Pflanze ab." The exact date of this film in unknown, but other films from the *Wunder der Natur* series were produced in the 1920s. All intertitles quotes come from notes taken during viewings at the state film archive in Berlin.

as in *Die Seele der Pflanzen* (*The Soul of Plants*): "Murderous desires slumber in the soul of the plant" ("Mordgelüste schlummern in der Seele der Pflanze!," *Die Seele der Pflanzen*, censorship card, Filmarchiv-Bundesarchiv). While human violence is not absent from these films—*Die Seele der Pflanze* includes one image of a scientist's hand burning a *Mimosa* leaf—such *Naturfilme* tend to gloss over violence in favour of a scientific picture of human relationships with plants.[161] The second act of *Das Blumenwunder*, however, directly thematizes the violence of science; indeed, its intertitles uncover the seemingly harmless didactic demonstrations in *Naturfilme* to reveal an essentially violent relationship, which technology—and in this case also film—perpetrates. Noticeably absent from this act are the flower dances, as *Naturfilme* are less about communicating with plants, reinforcing a sense of scientific mastery over the object.

In the second act, the violence of science towards plants echoes the violence of desire and greed thematized in the first act. The composition of the images in Act Two is reminiscent, in particular, of Karl Blossfeldt's well-known plant photography and connotes the same cool objectivity. But whereas Blossfeldt's images lack evidence of his manipulation of the plants and photos,[162] *Das Blumenwunder* shows hands and other tools entering into the frame and experimenting with the plants. Characteristic of such intrusions, one image shows two hands uncurling a vine from a stick, curling it in the opposite direction and pinning it to the stick (Figure 18). In other *Naturfilme* in the 1920s and 30s, the disembodied hands manipulate the behaviour of the plants, demonstrating the heliotropic striving of a plant upwards by turning the plant sideways in *Wunder der Natur* or the sensitivity of the *Mimosa* by burning it in *Geheimnisse im Pflanzenleben*. Cutting the scientist's face and its expressive subjectivity out of the frame, Act Two of *Das Blumenwunder* replicates in the hands the scientist's instrumental approach and highlights a relationship confined to the bare aesthetics of the laboratory. There, the relationship is of one

161 For a discussion of the disembodied hand as an emblem of the cinema of scientific vernacularization, see Oliver Gaycken. "'A Drama Unites Them in a Fight to the Death': Some Remarks on the Flourishing of a Cinema of Scientific Vernacularization in France, 1909–1914." (*Historical Journal of Film, Radio, and Television*. 22.3 (2002): 353–74. *Tayler Francis Online*, Web, 4 April 2014).

162 Part of the appeal of Blossfeldt's photos consist of the impression of objectivity they leave on the viewer. According to a collection of his working prints, it soon becomes apparent that his photos did not just reveal the plant as it is but magnified. Rather, the plants and the first images had been cleaned up to render a better or even ideal example of the plant. See Ulrike M. Stump, introduction to Karl Blossfeldt: *Working Collages*, ed. Ann Wilde and Jürgen Wilde (Cambridge, Mass.: MIT Press, 2001), 12.

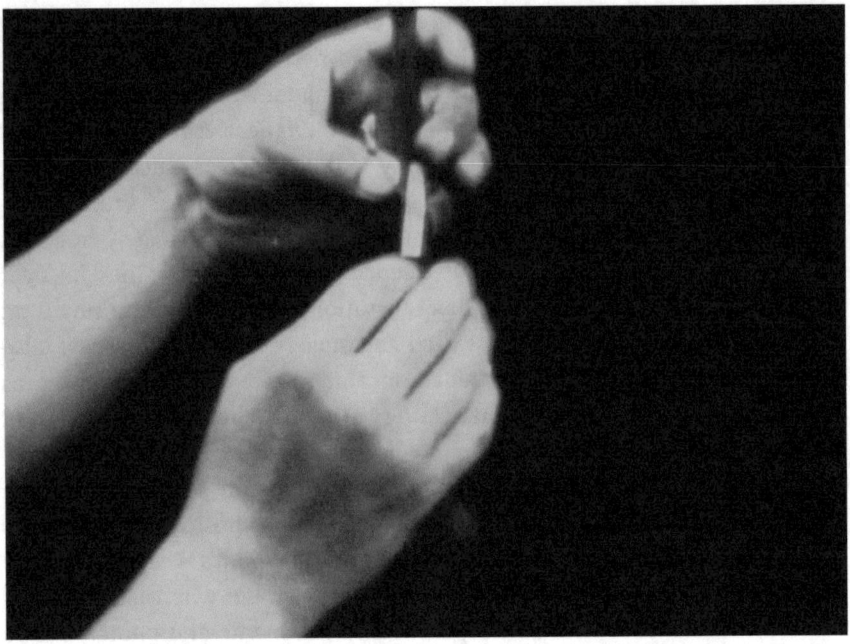

FIGURE 18 *Hands Manipulating a Plant from* Das Blumenwunder. *Scene still.* BASF AG, *Unterrichtsfilm, Verlag wissenschaftlicher Filme, 1922–26. Illustration reproduction courtesy of* Filmarchiv-Bundesarchiv.

of mastery, where an objectifying scientific gaze subdues nature, and where measuring and recording are emphasized in place of communication and negotiation. While the result of this relationship, violence, is similar to that resulting from the girls' desire for possession in the opening act, it is now impossible to identify with the hands. Even as a will is recognized in the plant through such descriptors as "uncertain" ("verzweifelt") and "struggle" ("Kämpfen") and in the uncanny human-like movements, its position as object prevents identification.[163] Yet, in the process of demonstrating a plant will—one that manifests itself precisely in its resistance to the intentions of the scientists—the possibility is opened up for a different interaction between humans and plants in the subsequent acts.

163 Produced over four years, the original purpose of the film, to advertise fertilizer—a means of manipulating the rate of plant growth, further reinforces the plant as an instrumentalized object.

It is here, in the third to fifth acts, that an alternative comes into view as conceptual language and rationalized nature are replaced in favour of communication through embodied language and through empathy. There are no longer any intertitles to explain and determine the plant images, which now centre almost exclusively on flowers; instead conceptual language is replaced by the language of dance. In their performance, the dancers mirror, translate and transform the time-lapse flower images, expressing the inner pathos of plants garnered through kinetic empathy. Other images—such as a man talking and a snake—reinforce their depiction of unity among plants, humans and animals. Without the disembodied hands and the mediating subtitles of the first and second acts, the viewer is now free to identify first with the dancers and then—through them—with the flowers. The tools of association, learned in the first act, and a sense of the plant's free will from the second act, empower the viewer to understand the non-narrative form in the last three acts as an alternative mode of communication with nature and with the image.

That this communication occurs through dance is hardly insignificant, for modern dance—and *Ausdruckstanz* in particular—was largely driven by the same desire to articulate an alternative to conceptual language that informed film theories such as that of Béla Balázs. Rudolf von Laban, one of the most influential theorists of *Ausdruckstanz*, largely understood his dances as a means of reconnecting bodily with the invisible rhythmical movements of nature—plants, crystals and the cosmos—and communicating such hidden 'life' to the spectator: "What is being awoken in the viewer [...] through the resonance with the actual dance piece,"[164] he writes in *Gymnastik und Tanz* (1925), "is [...] the animating of some power that puts us in the position to resonate together with the eternal primal universal dance of being" (16).[165] Laban's most famous student, Mary Wigman, defined modern dance for her part as a means of transforming knowledge (Wissen) into experience (Erlebnis):

> There, where the knowledge of things stop, where only experience is law, there, the dance begins. We don't dance feelings! They are already too definite, too clear. We dance the changes of the state of the soul, how it is

164 "Was [...]durch das Mitschwingen mit dem wirklichen Tanzkunstwerk im Zuschauer geweckt wird."

165 "ist [...] die Belebung jener Kraft, die uns instande setzt, mit dem urgesetzlichen Alltanz des Seins dauernd zusammenzuschwingen."

On this point, see also Evelyn Doerr, *Rudolf Laban. The Dancer of the Crystal* (Lanham: Scarecrow Press, 2008) (58).

> carried out in every instant in a particular manner and in the language of dance, it becomes a reflection of a human, an unmediated symbol of the entirety of living beings. (qtd in Müller, *jeder Mensch ist ein Tänzer*: 35).[166]

Several points in Wigman's definition shed light on the function of modern dance in *Das Blumenwunder*. Wigman identifies the boundaries of dance as a dynamic, individual experience rather than a cerebral understanding of the world. Like Laban, she saw dance as a means of communicating the unity of all living being. Nor was this view of dance as a means of bodily 'participation' limited to the theorists of *Ausdruckstanz*. Even a proponent of machine dance such as Oskar Schlemmer could figure dance as a means of empathetic communication between the dancer and his environment, one in which the dancer takes in the surrounding space and then responds from his innermost space (Müller 38).[167]

The concept of dance as a means of generating empathetic communication had its intellectual roots above all in the thought of Friedrich Nietzsche, in particular his championing of the "Dionysian" as an alternative to the rational relations favoured by Apollonian culture. One of the first theoreticians of modern dance, Isadora Duncan, largely crafted her vision for dance as an affirmation of life in unity with nature from her readings of Nietzsche's *Birth of Tragedy* and *Thus Spake Zarathustra*.[168] Upon closer examination of Nietzsche's *Birth of Tragedy*, many elements central to his understanding of dance became crucial to the interpretation of dance in *Das Blumenwunder*. In the first half of the book, Nietzsche begins by privileging dance as a means toward "the highest intensification of his [man's] symbolic powers," which

---

166 "Dort, wo das Wissen um die Dinge aufhört, wo nur das Erlebnis Gesetz ist, dort beginnt der Tanz. Nicht Gefühle tanzen wir! Sie sind schon viel zu fest umrissen, zu deutlich. Den Wandel und Wechsel seelischer Zustände tanzen wir, wie er sich in jedem Einzelnen auf seine besondere Art vollzieht und in der Sprache des Tanzes zum Spiegel des Menschen, zum unmittelbarsten Symbol alles lebendigen Seins wird."

167 Müller et al. interpret Oskar Müller's approach to dance as the following: "Der Tänzer war ihm die vollkommene Begegnung von Mensch und Raum, der 'Tänzermensch', wie er ihn nannte, nahm den ihn umgebenen Raum in sich auf und wirkte aus seinem Innern heraus auf ihn zurück" ("The dancer was to him the complete meeting of human and space, the dancing human, as he named him, absorbs into him the surrounding space and influences him in turn from within," 38).

168 Kimere Lamothe in Nietzsche's Dancers: Isadora Duncan, Martha Graham, and the reevaluation of Christian Values solidly connects Duncan's writings and specifically her use of the term Dionysian to Nietzsche's use of dance metaphor in *Birth of Tragedy* and *Zarathustra* (113–4).

can only be stimulated through a Dionysian aesthetic (21). Unlike the word or image in an Apollonian aesthetic, dance engages the entire body, which can respond to the Dionysian demands to express the "one-ness as the genius of humankind, indeed of nature itself" (21). The chorus, who danced and sang the Dionysian dithyramb in an Attic tragedy, was a medium, with which the spectator identified. Nietzsche described this moment of identification as a transmission of the feelings of the chorus to the crowd:

> Dionysian excitement is able to transmit to an entire mass of people this artistic gift of seeing themselves surrounded by just such a crowd of spirits with which they know themselves to be inwardly at one. This process of the tragic chorus is the original phenomenon of drama—this experience of seeing oneself transformed before one's eyes and acting as if one had really entered another body, another character. (43)

Nietzsche is illustrating through the Attic tragedy a spectatorial experience similar to that of kinetic empathy, when faced with a group of dancers. This experience lends itself to a feeling of connection with fellow spectators, with nature and with the primordial being. In essence, he is describing an embodied language, and privileging it over conceptual language, as do the final three acts of *Das Blumenwunder*.

Indeed, the dances in *Das Blumenwunder* clearly took part in this broader discursive context. Daisy Spies, the dancer who imitates the hyacinth blooming in the film, was an instructor and choreographer at the Mary Wigman School of dance, who also performed in Oskar Schlemmer's "Triadic Ballet." Max Terpis, who performs an abstracted rendering of a flower blooming and wilting, was at the time of the film the director of the Berlin State Opera Ballet and a former student of Laban. It was under his direction that the Berlin ballet corps performed in *Das Blumenwunder* an unusual flower dance far from the tenets of classical ballet.[169]

As dance comes to define the film during the last three acts, there is also a clear progression leading from mimesis of the natural world through dance to haptic perception of plants and dance and lastly, to a call to the viewers to actively participate. Distinguished from the second act by an intertitle, the third act initially leaves viewers to meditate on a series of time-lapse images of blossoming flowers, which express their creative will in the act of

169 Blankenship interprets this group dance as representative of Max Terpis's retreat from sheerly beautiful imitation of harmonic natural forms and also of a tendency within modern dance at this time to also reflect spiritual distortion (101).

unfolding, now unhindered by an interfering, disembodied hand. This meditation is broken only by an image of a snake and two dances, the hyacinth with Daisy Spies, and an unidentified flower dance with Elizabeth Grube. In both dances and the comparison of the snake, the similarities in form are highlighted as dissolves allow their shapes and movements to be momentarily matched with the flowers. Interrupting the flower images in a manner similar to the dancers, the movement of the snake is graphically matched with a root seeking nourishment and conveys the plant's connection to the animal world. Just as Nietzsche, Laban and others believed, movement here serves to reunite people with the natural world that includes the gestures of animals. Conditioned by Flora to see similarities in images, the viewer comprehends analogies between the dancer's movements, the snake and the plants. Furthermore, the art and will of the flower blooming resembles the will of the snake slithering on its way and the dancer, attempting to understand the flower through dance.

The first dance, the hyacinth by Daisy Spies, is a particularly powerful moment of mimesis designed to convey a unity of expression with the natural world (Figure 19). Spliced into the middle of a hyacinth blooming, Spies' dance appears seamlessly to continue the flower's creative act of opening in addition to matching its movement and form. Just as the flower performs in the film studio with a black background, so Spies performs on stage against a dark background. In dance as well as through the film form, her movements are chosen to mirror nature. Her hands, bent sharply at the wrist, echo the hard angles of the hyacinth's petals just as each of her subsequent gestures closely follows the movements of the flower blossoming. In place of the antagonistic relationship embodied by the scientist's hand in the previous act, the film now shows us the productive power of the earth and the stage, where both the dancer and the flower resist objectification. By 'becoming flower' on the screen, Spies thus recalls Walter Benjamin's "Doctrine of the Similar", according to which, "[t]he gift which we possess of seeing similarity is nothing but a weak rudiment of the formerly powerful compulsion to become similar and also to behave mimetically" (69). Seen within this context, Spies' dance, in the midst of the hyacinth blooming, initiates in the viewer the drive to see similarities between the creativity of dance and the flower blooming, and thus awaken memories of an experiential mode in which nature was perceived as having its own 'life' or subjectivity.

However, the dancers' response to the gestures of flowers in the film is not limited to simply parroting the plant, but rather seeks to transpose the plant's movements into a new bodily language. Transposing nature's forms means retaining the sensuous and recognizable link to nature, instead of abstracting from the physical experience of the natural world like conceptual language. While the fourth act also foregoes explanatory titles and intertitles, the shift

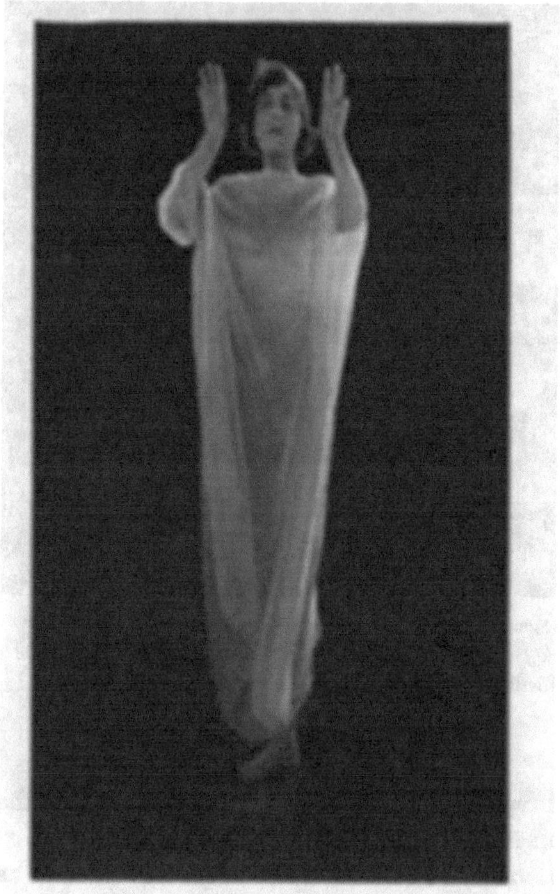

FIGURE 19 *Hyazinthe, Daisy Spieß from Souvenir Program, 1926.* Archiv der Deutschen Kinemathek. Arte.tv. *Web. 18 Nov 2013. Illustration reproduction courtesy of the* Archiv der Deutschen Kinemathek.

from mimesis and unity in nature to bodily language distinguishes it from the third. In this section, three dances by the Berlin state opera, Max Terpis and Stefa Kraljewa as well an image of man speaking and waving his hands, seek to express the experience of being a plant beyond imitating the movement made visible through film. With the help of film, the visual analogies between the dancers and the flowers reveal the expressivity of plant movement, leading to the rudiments of a bodily language. Occurring part way through the act, the interaction of the image of an excited man gesturing while speaking with

FIGURE 20 *Dancers Leaping in Slow Motion from* Das Blumenwunder. *Scene still.* *BASF AG, Unterrichtsfilm, Verlag wissenschaftlicher Filme, 1922–26. Illustration reproduction courtesy of* Filmarchiv-Bundesarchiv.

the subsequent image of a twitchy plant growing is one example of how the emotional register of gesture is transferred to the images of plants.

The potential combination of dance and film as bodily language finds its fullest expression in the group dance by the Berlin state opera ballet corps. In their ballet, the group dancers transpose the flower movements in a manner similar to Max Terpis and Stefa Kraljewa, but here the sensual effect is compounded through film techniques and the extended possibilities of a group of dancers. Instead of remaining in one position like the rooted plant (as did other dancers), the ballet dancers extrapolate plant gestures into forming rings and other patterns seen in the structure and movement of plants. As a group of dancers moving together, they also simulate the interaction of plants with one another and with different parts of themselves. Similarly, slow motion takes of a dancer leaping (Figure 20) and a medium shot of a dancer twirling surrounded by others further breaks down the flowing dynamic of the natural world into comprehensible parts.[170] Similar to the chorus in Greek tragedy, the group dance also permits spectators to place themselves amongst the dancers

170 Blankenship interprets the shift from fluid movement to slow-motion heaviness as indicative of the Zeitgeist and of Max Terpis' aesthetic inclinations towards ugliness (101). This

and experience through kinetic empathy unity within an ecology. In the context of the film, the use of slow motion shows the heaviness of the slow struggle upwards of the plant and its rootedness in the earth. The slow motion also creates a sense of the plant's heaviness in the viewer. The heliotropic striving of a plant and the effort it takes to blossom is felt by the viewer as heaviness. Thus, film technique is implicated in simulating the experience of plants as both a language of movement and as an experience of that movement. The haptic visuality of film and the kinetic empathy of dance combine to create a visual aesthetic envisioned by Balázs.[171]

This bodily language then provides the basis for the film's final appeal to the spectatorship in the fifth and final part, where a combination of haptic and optical visuality induces the viewers to project themselves onto the screen and translate for themselves the flower movements. Unlike the previous two sections, the last act returns in a modified way to conceptual language by beginning with an intertitle, "The Song of Becoming and Passing Away" ("Das Lied vom Werden und Vergehen"). As compared to the intertitles in the first and second act, however, language here draws on the discourses of music as the language of the soul and in no way diminishes the embodied nature of the following images. Beginning the act, a dancer, depicting the life and death of a flower, performs one last example of bodily communication with plants for the viewer. Dancer Herbert Haskel expresses the death throes of a plant through passionate arm movements as he crumples to the floor. The subsequent flower images are no longer interspersed with dances or other image clues, but rather only with images of the garden—now empty of other people. If the return of the garden reminds the viewer of the unnatural death of the flowers at the hands of the girls from the film's opening, it also suggests a blank screen. The empty rows of flowers, that is, are like the blank spaces

---

appearance of heaviness, she describes, could also be indicative of Terpis' wish to extricate his choreography from the ethereal lightness of classical ballet.

171 A similar moment occurs in another popular Weimar film, *Wege zur Kraft und Schönheit* (1925) with a group of dancers. The film techniques express an affinity with nature, however, differ from Das Blumenwunder with respect to the level of overt preparation the audience receives and the integration of dancers and nature. At the beginning of the fourth section, titled "The Dance" ("Der Tanz"), dancers are superimposed upon a forest scene with yet more dancers sitting in the trees and dancing in a circle far in the background. The dancers' movements present a striking likeness to the waving branches of the trees to the point where it becomes difficult to discern where their bodies end and where the trees begin. Like in the earlier images from the Loheland school, dancers are presented as mirroring the natural rhythms, but here the film technique unifies the dancers with nature. Like the dancers from the Berlin state opera, an embodied language is not just confined to the bodily gestures, but also encompasses film gestures.

in an exercise sheet, into which the viewer can project herself. The kinetic empathy the viewer experienced through the paired gestures of flowers and dancers in the earlier scenes now translates into an aesthetic practice of empathy where the viewer can apply what she has learned in the absence of intermediary figures.[172]

The images of flowers blossoming and wilting in the last act are meant to evoke in the viewer the feelings of a plant struggling to live and die. There is evidence that time-lapse images of plants from the 1920s evoked this kind of physical sensation in their audiences, beyond the sensation of touch provoked by images of textures already recognized by Marks and Sobchacks as haptic visuality. As French author Sidonie-Gabrielle Colette recalls from screenings in the 1920s, the animal-like movement of plants projected in time-lapse were felt so intensely by children in the audience that they were then compelled by the sensation to copy the plant movement:

> A "fast motion" documentary documented the germination of a bean [...] At the revelation of the intentional and intelligent movement of the plant, I saw children get up, imitate the extraordinary ascent of a plant climbing in a spiral, avoiding an obstacle, groping over its trellis: "It's looking for something! It's looking for something!" cried a little boy, profoundly affected. He dreamt of a plant that night, and so did I. These spectacles are never forgotten...[173] (qtd in Blankenship: 88, 61)

In imitating the plant's movements, the children recognize the bodily language as meaningful and translate the movement into conceptual language. The viewing experience described by Colette is a complex mix of physical and conceptual realities barely indistinguishable from one another. The

---

172 Such intermediary figures were not limited to bodies on the screen. At the premier of *Das Blumenwunder* in Breslau on June 4th 1926, the screening was followed by Ursel-Renate Hirt's performance of three flower dances set to music by Strauss, Poldini and Gounod and inspired by nature poetry by Goethe and Felix Dahn. Her addition to the program is reminiscent of Nietzsche's Dionysian aesthetic, which privileges poetry, music and the body as the gates to a unified experience with nature. Like the chorus in the Attic tragedy, she has the potential to stimulate the audience to feel the psyche, the soul and the humour of flowers. Through dance she reinforces the creative will inherent in an embodied language as it also appears in nature.

173 I first came across this description in Blankenship's article on Das Blumenwunder. Colette's observation proved to be the seed for my reading of this film as particularly geared for interactive viewing. It is clear that the time-lapse images of plants growing were not just seen by the children but also felt.

switching between optical visuality and haptic visuality in the last Act of *Das Blumenwunder* expresses the complexity of the film viewing experience.

In its progression from articulated language to mimetic understanding of plant life and ultimately a language of the body, *Das Blumenwunder* rehearses an interrogation of conceptual language and its ability to convey authentic experience that was part and parcel of modern European aesthetics. As Maurice Maeterlinck explained it already twenty years before *Das Blumenwunder* in his *Treasure of the Humble*: "How strangely do we diminish a thing as soon as we try to express it in words! We believe we have dived down to the most unfathomable depths, and when we reappear on the surface, the drop of water that glistens on our trembling finger-tips no longer resembles the sea from which it came" (77). For symbolist poets such as Maeterlinck, the act of writing involved a constant struggle against language—its automatisms, its conceptual categories and its syntactical parceling of reality into grammatical objects and subjects—in order to convey an experience lying just beyond the limits of language's grasp. Within this context, visual media such as dance and film began to hold out a particular appeal as alternatives for conveying non-linguistic experience. Thus it should come as no surprise that writers such as Hugo Hofmannsthal—who penned perhaps the most compelling account of the crisis of conceptual language in his *Ein Brief* (*A Letter*, 1901)—would turn to these media in works such as *Das fremde Mädchen* (*The Strange Girl*), Hofmannthal's 1911 pantomime that was made into a film with the dancer Grete Wiesenthal in 1913.

Taking up this struggle with articulated language, *Das Blumenwunder* stages the gradual elimination of titles to suggest that a dynamic and embodied language is better suited to the experience of "plant life," even as the film acknowledges that the experience it offers is still a mediated one. By returning as close as possible to a direct experience of plants through mimesis, the film also suggests that all forms of mediation from art to language to even technology are rooted in the natural world. Understanding ourselves then would begin by returning to this initial moment of mimesis. This film asks its viewers not only to see the similarities on the screen but then to dance these movements in an effort to understand the miracle of life—the experience of a flower as it blooms and wilts—in a mode going beyond the conceptual grasp of scientific knowledge.

A travelling exhibition (2002–2011) designed by the Science Museum of Minnesota, *Playing with Time*, takes the interactive engagement of films like *Das Blumenwunder* a step further. One station of the exhibit invites the museum visitors to step into time-lapse images of growing plants through the use of green screen technology and to imitate the movements of the plants.

FIGURE 21 *Dancing with Plants from Science Museum of Minnesota: St. Paul. Web. 6 Jan 2014. Illustration reproduction courtesy of the* Science Museum of Minnesota.

The visitors can then view themselves "dancing" with the plants projected onto a screen (Figure 21). The exhibition creates a space for empathetic engagement with plants where visual clues and physical imitation occur simultaneously. Unlike the viewers in the 1920s, who had to be explicitly taught the meaning of the time-lapse images of plants through a mediator, Flora, the young attendees understand immediately the plant movement as communication through gestures, a dance that enables embodied empathy. The film, *Das Blumenwunder*, and the exhibition point to the significant and often overlooked contribution of the arts and media to our knowledge of nature: they can engage with one of our basic mechanisms of learning and connecting with others socially, our body.

CHAPTER 4

The Radical Other: The Metamorphosis of Humans and Animals into Plants in Gustav Meyrink's "Die Pflanzen des Doktor Cinderella" (1905)

> And everything appeared as parts, taken from living bodies, put together with an inconceivable art, robbed of their human soul and suppressed down into pure vegetative growth.
>
> "Die Pflanzen des Doktor Cinderella"

∴

This chapter marks a shift in focus from uplifting and utopian portrayals of plants to plants as monsters (Figure 22). Since Darwin introduced insectivorous plants to the broader public in 1875,[174] their unusual behaviour quickly drew attention and piqued the imagination of writers, filmmakers and artists, inspiring many representations of these plants that emphasize their deviant and monstrous qualities. The popular name for insect-eating plants, carnivorous or meat-eating plants, reflects the resonances these plants had in the public imagination as predators.[175] As one of the foremost specialists in esoteric and fantastic fiction, Gustav Meyrink seems to have been particularly drawn to the combination of plants as predators, returning to the vampiric plant in at least three short stories. One of these stories, "Die Pflanzen des Doktor Cinderella" ("The Plants of Doctor Cinderella," 1905),[176] imagines

174 Charles Darwin published his book on insectivorous plants, *Insectivorous Plants*, in 1875 with an American and British printing, followed by a second edition in 1888. The English version was swiftly followed by a German translation, *Insektenfressende Pflanzen* in 1876 and a French translation, *Les plantes insectivores*, in 1877.

175 In German, these plants are called either "Fleischfressende Pflanzen" or "Karnivoren." As an English native speaker, the first name recalls for me the English word "flesh" even though the word technically means "meat", giving the name a particularly grotesque character.

176 The short story was first published in Simplicissimus in 1905. Later, the story was included in a collection of short stories called Wachsfigurenkabinett: Sonderbare Geschichten. (Wax Figure Cabinet: Strange Stories),

 | DOI 10.1163/9789004327177_006

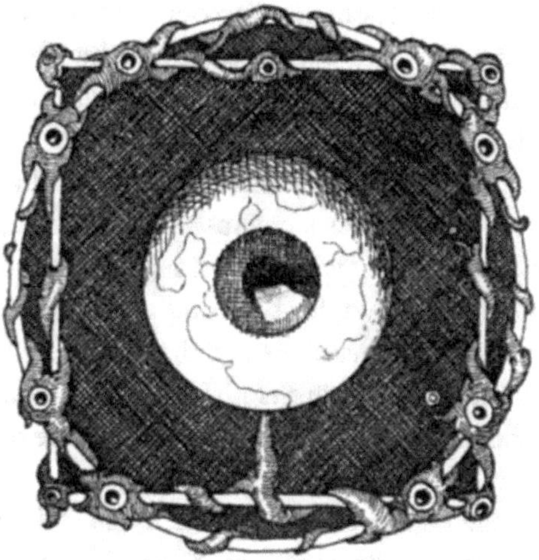

FIGURE 22
Eye Illustration from André Lambert, "Die Pflanzen des Doktor Cinderella," Wachsfigurenkabinett *by Gustav Meyrink. München: Albert Langen, 1908. 203.* Internet Archive. *Web. 17 Jan 2013.*

carnivorous plants as blood-sucking, meat-eating monsters without any physical resemblance to the insectivorous plants Darwin found so intriguing. Instead, his carnivorous plants are composed of human and animals body parts that have been reduced to the base function of plant growth. In this chapter, I argue that Meyrink's complex story imagines the grotesque and monstrous plants as a consequence of two seemingly disparate approaches to understanding life, the occult and medical sciences, which share a Cartesian dualism of mind and body.

Some of the challenges to interpreting Meyrink's story lie in his interweaving of science and the occult, and the resulting fragmented narrative and narrator. The narrative deliberately misleads the expectations of the reader, before eventually revealing that the narrator and the evil scientist "Dr. Cinderella" of the story's title are in fact one and the same person. The narrator, plagued by mental blanks,[177] tells a murky story of how he came to be in his current situation, mentally and physically crippled. He blames it on a statue, "the bronze" ("die Bronze"), that he accidentally found buried in the sand while in Thebes, Egypt. The statue is a likeness of the ancient Egyptian god, Anubis, in a particular pose that is said to imitate a hieroglyph. After finding the statue and returning to Europe, the narrator is caught by an incredible curiosity that insatiably drives him to unlock the statue's secret. After imitating the statue's pose, he begins to have periods of lost time—suffers from mental blackouts—and has visions of mysterious, phantasmic figures. One night, while wandering the

177 The narrator calls this "geistesabwesend" which literally means absent minded.

streets, he is drawn to a house and enters it. There, he discovers a grotesque laboratory filled with carnivorous plants that had been pieced-together from animal and human body parts: "What I ever felt of fear and horror was nothing in comparison to this glimpse. [...] The wall was covered with a net of tendrils—blood red arteries, out of which swelled hundreds of staring eyes like berries (82).[178] In horror, he rushes out of the house, collapses and is helped by a police officer to the commissioner's office, where he learns his name, Dr. Cinderella, and that he is known around town for breeding new types of carnivorous plants, the Nepenthes and the Drosera.[179] After that night, he is never again able to find the house with the carnivorous plants, no one from the commissioner's office knows of the incident, and he has been left lame on one side of his body.

The type of carnivorous plants at the centre of "Die Pflanzen des Doktor Cinderella" seem to have formed a recurrent motif in Meyrink's work. Vampirism and plants are first brought together in an earlier story by Meyrink, "Bologneser Tränen" ("Bolognese Tears"), published in a short story collection from 1903. The title of the collection, *Orchideen: sonderbare Geschichten* (*Orchids: Strange Stories*), refers to the main female character, Mercedes, from "Bologneser Tränen." She briefly transforms into an orchid: "In this moment, let loose from the dark, tangle of leaves, rose up a huge orchid,—the face of a demon with longing, thirsty lips,—without a chin, only dazzling eyes and a gaping, bluish seam. This plant face shook on its stalk, swaying in wicked laughter and staring at Mercedes' hands [...] "Do you believe that orchids can think?" (80).[180] The sexualized and demonic description of an orchid blossom attributes Mercedes' human characteristics to a plant, while the subsequent

178 "Was ich je an Furcht und Grauen empfunden, war nichts gegen diesen Augenblick. [...] Mit einem Rankennetz blutroter Aldern, aus dem wie Beeren hunderte von glotzenden Augen hervorquollen, war die Mauer bis zur Decke überzogen."

179 The Nepenthes and the Drosera refer to actual carnivorous plants, which digest insects. The Nepenthes is commonly called a pitcher plant, because it lures insects to the slippery edge of its trap that resembles a pitcher. The insects then fall in and are digested by the plant's juices. The Drosera is commonly known as the Sundew because of sticky and sweet-smelling drops at the ends of stalks that resemble dew. Insects that brush against these stalks are caught, gradually enveloped by the stalks and digested. The plants described in Meyrink's story in no way resemble these two carnivorous plants.

180 "In diesem Augenblick schnellte, losgerankt aus dem Dunkel des Blättergewirres, eine riesige Orchidee,—das Gesicht eines Dämons, mit begehrlichen durstigen Lefzen,—ohne Kinn, nur schillernde Augen und ein klaffender, bläulicher Saumen. Dieses Pflanzengesicht zitterte auf seinem Stengel; wiegte sich wie in bösem Lachen, auf Mercedes Hände starrend. [...] Glauben Sie, daß Orchideen denken können?"

description gives her vague plant-like qualities: "And she was an orchid queen, this Creole woman with her sensuous, red lips, the soft, greenish shimmer of her skin and her hair the colour of dead copper" (80).[181] She vacillates between a human-like plant and a plant-like human whose seductive blossom-face masks the deadly threat of her snake-like body hiding underneath. Her attraction reflects the elaborate beauty used by an orchid in nature to attract insects in order to pollinate, and in turn, frames the orchid's charms as vampiric.[182] In this story, Meyrink blurs the boundary between human and plant, exposing the threat powerful and deadly female sexuality poses to masculine power and normalized gender roles. The comparison of a dangerous plant with the threat of female sexuality is a common one and more thoroughly examined in chapter five with the film *Alraune* (1928).

The intersection of plants and vampires reoccurs in a third story by Meyrink, "Der Kardinal Napellus" ("The Cardinal Napellus"), as a warning against false religions. First published as a part of a short story collection, *Fledermäuse* (*Bats*) in 1916, the story follows the experience of a monk, Radspieler, when he joins a group of monks. Known as the Blue Brothers (Blaue Brüder), the monks breed a plant with blue flowers referred to as monkshood as well as the Latin, *Aconitum napellus*, which causes a trance-like state giving the appearance of death ("Scheintod"). The plant in the story refers to an actual poisonous plant with the same name, giving the story some basis in the botanical sciences. Each monk is attached to a particular plant, which lives only through ingesting the monk's blood in a garden adjacent to the cloister: "Behind the walls of the cloister is a garden, in which a flower bed blooms in summer full of those deadly herbs, and the monks water them with blood that flows from their flagellation wounds. Each one, if he is a brother of the community, has such a flower to plant, that then, like in baptism, receives his own Christian name."[183] Gradually, it becomes apparent to Radspieler that the poison produced by the plants does not lead to a greater spiritual enlightenment but rather to death

181 "Und sie war eine Orchideenkönigin, diese Kreolin mit ihrer sinnlichen, roten Lippen, dem liese grünlichen Hautschimmer und dem Haar von der Farbe toten Kupfers."

182 Several orchid species mimic the appearance of a receptive female insect, tricking the male into mating with the flower. For example, the Catasetum orchid gives off the an attractive scent for male bees, who collect it. Male flowers have the added effect of "violently" attaching a pollen packet to the back of bees, causing the bees to avoid the plant in the future (Alcock 42).

183 "Hinter den inter den Klostermauern liegt ein Garten, darin blüht im Sommer ein Beet voll von jenem blauen Todeskraut, und die Mönche begießen es mit dem Blut, das aus ihren Geißelwunden fließt. Jeder hat, wenn er Bruder der Gemeinschaft wird, eine solche Blume zu pflanzen, die dann, wie in der Taufe, seinen eigenen christlichen Namen erhält."

and he leaves, seeking the mysteries of life in watery depths. The plant and his experience at the cloister continue to haunt him for the rest of his life and eventually drive him to insanity. Meyrink interweaves the romantic motif of the blue flower—well known from Novalis's novel *Heinrich von Oferdingen*—with the deadly biology of monkshood—ingesting the flower's poison and feeding the plant with their blood. Through Radspieler, Meyrink criticizes idol worship that calls for self-abnegation. Meyrink is also criticizing a sentimentalizing of nature that is one of the offshoots of the romanticizing of nature as harmonious and unifying.

Plants, vampires and the occult tie together "Bologneser Tränen," "Der Kardinal Napellus" and "Die Pflanzen des Doktor Cinderella." More precisely, as Amanda Boyd argues, all three stories point to the dangers of occult practices through the common figure of the vampire. In "Dr Cinderella's Plants" specifically, Boyd reads occult practices as responsible for the fragmenting of Dr. Cinderella's mind and body (608). Meyrink's fascination with the occult has been well documented and its role in his many short stories and novels, thoroughly examined. He belonged to many occult organizations that pursued with conviction a deeper meaning hidden beneath the surface, yet he maintained a vibrant scepticism and wit (Lachman 209). Meyrink extended the reach of his critical and satiric gaze beyond the confines of the occult to warn against other idols in society. The demonizing of plants common to all three stories is also intriguing for their role in qualifying his critique.[184] Meyrink recognizes the potential affective power of inverting the perception of plants as passive and harmless into predators for a horrifying effect. Additionally, his portrayal of plants as dynamic and active resonates with the changes in the perception of plants at the end of the 19th century.[185]

Plants as monsters and specifically vampires is by no means an arbitrary choice and reflects on the peculiar position of plants as alive yet clearly inhuman. Plants have been consistently placed on the hierarchy of being as closer to crystals and other seemingly inanimate forms because they are perceived

184 Other short stories, in which plants play a significant role, include: "Meister Leonard" (Master Leonard), "Die Urne von St. Urne von St. Gingolph" (The Urn of St Gingolph) and "Blamol."

185 After Darwin published in 1875 the second edition of *The Movements and Habits of Climbing Plants* and *Insectivorous Plants*, both subsequently translated into German, there followed a surge of interest in the movement of plants as examples of similarities between plants and animals, including sensitivity to the environment and animal-like movement. One such example comes from Viennese scholar, Alois Pokorny's, lecture from 1878 regarding the life of vines on Darwin's work.

to be missing the parts that represent animal and human life. Blood, breath, movement and a mind have long indicated life within Aristotelian and Judeo-Christian traditions of thought and give reason for plants to be set apart from animals and humans as inferior beings and soulless (Hall 61). In contrast, the shared faculties of animals and humans give rise to the perception of similarity. Matthew Hall, a scholar interested in the relationship between plants and humans, calls the plant in this role the radical other—perceived as being fully non-human.

When benign plants become blood- and fat-sucking monsters as in Meyrink's story, "Die Pflanzen des Doktor Cinderella," their position as the other is paradoxically emphasized and problematized, giving rise to the question of what it means to be human. Dr Cinderella has pieced together vines with berries from veins and eyes, mushrooms from fat and skin, and a pine cone from human fingernails. Their parts are identifiable as from humans and animals as exemplified in description of the pine cone as made "from rosy human fingernails!" ("Tannenzapfen aus rosigen Menschennägeln!," 84), and the discovery of a man's body with the "fingernails ripped out" ("die Nägel ausgerissen," 85). Similarly, the narrator's description of the eyes identifies a variety of human and animal sources: "Gruesome, countless eyeballs glittered in between,[186] which alternating with hideous, blackberry-like bulbs spurted up and slowly followed me with their gaze as I went by—eyes of all sizes and colours—from the clear-shimmering iris up until the light blue, dead, horse eye, which stood upward and unmoving" (83).[187] The body parts, while derived from animals and humans, in no way transform the plants into the familiars of humans or animals. Instead the narrator is physically repulsed by what he sees: "Every fibre of my body cried out in indescribable horror" ("Jeder Fiber meines Körpers brüllte auf in unbeschreiblichem Entsezten," 82). On the surface, the horrifying effect can be directly attributed to the transformation of ordinary plants into carnivorous ones in addition to the gruesome experience of seeing human bodies in pieces. Yet, the gruesome plants also stand for the subversion of the hierarchy of being that connects rational thought with the human body at the pinnacle. The image of human and animal body parts as plants frames predatory behaviour of carnivorous plants as perverse

186 Eyeball is closely connected to fruit in German, literally translated as eye-apple.

187 "Grausig glitzerten dazwischen zahlose Augäpfel, die in Abwechslung mit scheußlichen, brombeerartigen Knollen hervorsproßten und mir langsam mit den Blicken folgten, wie ich vorbeiging.—Augen aller Größen und Farben.—Von der klarschimmernden Iris bis zum hellblauen toten Pferdeauge, das unbeweglich aufwärts steht."

and unplant-like, while it brings out the aspects of animals and humans that are seen to be plant-like.

The animal-like movement of Dr. Cinderella's plants draws on discourses surrounding movement and plants that reconsider the distinction between plants and animals as well as the perception of life in plants. The narrator viscerally experiences the life in plants as animal-like: "The inconceivable was only that these plants, or what they otherwise might be, felt blood warm and brimming, and wholly made a fully animal impression on the sense of touch" (81).[188] As with animals, the plants' movement is a clear indication of life for the narrator: "That there was life in them, I clearly recognized, when I closer lit the eyes and saw how the pupils immediately tightened" ("Daß Leben in ihnen war, erkannte ich deutlich, wenn ich die Augen näher beleuchtete und sah, wie sich sofort die Pupillen zusammenzogen") (84). Movement is a reoccurring feature of Dr Cinderella's plants: The eyes have a "twitchy movement" ("zuckender Bewegung"), and the mushrooms made of fat and skin, "winced at every touch" ("bei jeder Berührung zusammenzuckten," 84). The animation of these plants recalls a common argument for plants as living beings popular in the mid-19th century and the *fin de siècle*: Gustav Fechner, Maurice Maeterlinck and Raoul Francé among others saw movement as characteristic of plants, often using discernibly sensitive plants like the *Mimosa pudica* and the many carnivorous plants as examples of plant sentience. Charles Darwin, in his book on climbing plants, refers to movement as one of the "vague" assertions of difference between plants and animal.[189] The many nature films from the early 20th century, including *Das Blumenwunder* made use of film's capacity to shorten time and to speed up the imperceptible movement of plants revealing the presence of an uncanny resemblance to animals. These scholars, writers and filmmakers frame plant movement as a part of the natural world and a point of similarity between animal, human and plant life. In contrast, Meyrink's story frames the animation of Dr. Cinderella's plants as the unnatural work of an anatomist and as the result of transferring human and animal life to plants and revealing the plant within the human. In the descriptions

188 "Das Unbegreifliche war nur, daß sich diese Pflanzen, oder was es sonst sein mochte, blutwarm und strotzend anfühlten und überhaupt einen ganz animalischen Eindruck auf den Tastsinn machten."

189 Darwin sees movement not as intrinsic to animals, but rather as a characteristic that has an evolutionary purpose: "It has often been vaguely asserted that plants are distinguished from animals by not having the power of movement. It should rather be said that plants acquire and display this power only when it is of some advantage to them." (206).

of Dr. Cinderella's laboratory, life appears in its most basic form, blood and movement, without preserving the soul.[190] Movement, rather than becoming a point of similarity, becomes a moment of difference.

The plants become monstrous because they blur the boundaries between humans and plants, while emphasizing difference. Scholar Jeffrey Cohen argues in his analysis of monsters that one of their main functions is to threaten the stability of identity through their challenge to categories of being: "And so the monster is dangerous, a form suspended between forms that threatens to smash distinctions" (6). they are disturbing Though the plants in Meyrink's story are composed of human and animal parts, they have robbed these parts of their humanity: "And everything appeared as parts, taken from living bodies, put together with an inconceivable art, robbed of their human soul and suppressed down into pure vegetative growth" ("Und alles schienen Theile, aus lebenden Körpern entnommen, mit unbegreiflicher Kunst zusammengefügt, ihrer menschlichen Beseelung beraubt, und auf rein vegetatives Wachstum heruntergedrückt," 84). The narrator perceives the plants as having been artificially constructed and reflects the assumption that plants do not have a soul, recalling Aristotle's tripartite soul that recognizes in plants only a nutritive soul as compared to the perceptive soul in animals and the rational soul of humans.[191] Through the animal and body parts, the plants have acquired animal life as represented in blood and movement, but are missing the soul as represented through the breath and mind. In the Judeo-Christian tradition, the soul has been commonly associated with the breath of life—at once linking the soul to the insubstantial, air, and to a movement not associated with plants (Hall 57).[192] The vivisection of human and animals

---

190 The plants are not the only supposedly inanimate objects that have come alive: "Aus dem Nebel taucht ein Haus—mit abgebrochenen Schultern und zurückweichender Stirn, und glotzt besinnungslos aus leeren Dachlocken zum Nachthimmel auf wie ein verendetes Tier" (78–79).

191 Aristotle defines the plant soul as: "Since nothing is nourished which does not partake of life, what is nourished will be the ensouled body insofar as it is ensouled, with the result that nourishment (i.e. food) is related to the ensouled, and not coincidentally" (qtd in Shields: np). He seems to mean that food distinguishes inanimate objects from animate. The nutritive soul is common to all living beings, yet just a step up from inanimate objects.

192 Breath also distinguishes Meyrink's monstrous plants from the dynamic living plants in Scheerbart's short story, "Flora Mohr: Eine Glasblumen-Novelle." While Scheerbart's plants do not breathe, their creator, Wilhelm Weller has attempted to breathe life into

and the reconstruction of those parts as plants makes a profound statement on what it means to be human, a tangled combination of body and spirit.

The eyes of the plants are a particularly powerful expression of humanity reduced to mere substance and vegetative drive. A wall of eyes is the first thing the narrator encounters in the laboratory, and is introduced with three separate qualifications of the horror that he experiences at the sight of those eyes (82). His horrified reaction is caused by the evidence of a sentient plant that responds to him but is missing a soul: "The one [an eye], which I just grabbed, still sprang in twitchy movement to and fro and squinted at me maliciously" ("Das eine, in das ich soeben gegriffen, schnellte noch in zuckender Bewegung hin und her und schielte mich bösartig an," 82). On account of their movement, the eyes are perceived as alive and aware by the narrator, but instead of offering a glimpse of a benign plant soul, they give the narrator the evil eye. While Dr. Cinderella has transferred animal and human life to plants, he has not succeeded in transferring the soul. Instead, the plants represent merely "vegetative growth." The narrator perceives human and animal eyes to be reduced to a reflection of the inner life of a plant as demonic and soulless.

The sewing of eyes onto a plant opens up one of the fundamental framing devices for plants that defines plants in accordance with what they do not have. For Aristotle, one of the essential absences in a plant are the eyes. He says, "the plant is said to be 'deprived' of eyes" (book v, chapter 22). In his book, *Plant-Thinking* (2013), Michael Marder sees the Aristotle's privileging of the plant's deprivation as an ambivalent point. He asks if it is "an irreparable flaw or a sign of the plant's quasi-divinity?" (23). He continues on to note that Aristotle's plant deprivation contrasts with Theophrastus' vision of the plant's vivacity in all its parts. As he concludes, it is Aristotle's model of deprivation that has received the most attention in Western philosophy over Theophrastus' "exuberance of vegetal life" (23). In Meyrink's short story, Aristotle's model of deprivation is confirmed by the presence of the eyes and the accompanying exuberant animal life pulsing through the plant. The eyes offer such a potent symbol of the plant's soulless, demonic nature, precisely because their presence reaffirms the original deprivation as a sign of absence.

The significance of the eyes for the presence of consciousness belongs to a series of differences between plants, animals and humans during the 19th century that are interpreted as signs of a plant soul. The differing relationship of the parts to the whole between plants and animals raises fundamental questions as to the integrity of the individual whole. On the surface, it appears

---

them and therefore, give them a soul. See chapter one page 17 for a discussion regarding the role of Weller as the creator of living, ensouled plants.

that in animals and humans, parts cannot be added nor subtracted without disturbing the integrity of the whole. In contrast, plants have been framed as in an incomplete state, constantly supplementing their form with new parts and subtracting old ones. This incomplete state has been used as a basis to argue that plants are neither sentient nor do they possess a soul. Eduard Schmidlin's book on popular botany from 1867 is characteristic of this argument, associating their type of movement and the state of completion with a soul. He writes:

> The changes in growth and shape occur in the animal through the transformation of the old into the new, not as with the plant through the adding of the new to the old, and from this different way of growing [...] Out of these differences in the development of form in animals and plants comes the distinct capacity of animal development—the ability to sense and initiate movement, which forms the basis for the activity of the soul. [...] A life, however, that never is a complete whole, but rather continues in a succession of parts, that according to their formation, without renewing themselves, die again, like the plant, can be capable of neither awareness nor self-initiated movement. (7)[193]

Schmidlin's key point hinges on the concept of awareness and reveals his reliance on a zoomorphic concept of a sense of self. He assumes that as with animals or humans, losing a body part would be traumatizing and the absence of any evidence of trauma in plants is itself evidence of the absent awareness.

Part of what makes the pieced together plants in Meyrink's short story monstrous comes from applying this principle of incompletion to the human body. Once removed from the humans and the animals, the body parts take on a new vegetal meaning. They have been transformed from parts that are subservient to the organism as a whole and signal its completion. Instead, the individual parts are dispensable—they die off and can be grown again without changing the

193 "Das Wachsthum und die Gestaltveränderung geschehen also bei dem Thiere durch Verwandlung des Alten in das Neue, nicht wie bei der pflanze durch Hinzufügen des Neuen zum Alten [...] Auf diesem einschneidenden Unterschiede in der Formentwicklung beruht endlich auch die Fähigkeit der thierischen Bildung, die Unterlage seelischer Thätigeiten, der Empfindung und selbsthätigen Bewegung, zu sein. [...] Ein Leben abr, das nie ein abgeschlossenes Ganze ist, sondern nur in einer Reihe von Theilen verlauft, die nach ihrer Bildung, ohne sich zu erneuern, wieder erstarren, wie eben die Pflanze, kann weder der Empfindung noch der selbsthätigen Bewegung fähig sein."

status of the plant from complete to incomplete. The difference lies in the meaning that the eyes hold for the person they once belonged to and the new meaning that they take on for the vine. For this reason, the loss of the eye for the person is significant but the loss of an eye (the berries) for the vine is both inevitable and insignificant. The image of the dead horse-eye in Meyrink's story powerfully represents this new vegetal meaning. It points to the part's original meaning as complete organism that as a whole organism was a vessel for a soul. In its new meaning as a berry, its death also denies the plant the wholeness needed for the organism to contain a soul. The dead horse-eye also indicates a second more fundamental, ontological ambivalence to plants that places them halfway between life and death. They add new living parts, while old parts die and drop off. The pieces, which can never form a complete whole, also cannot provide the basis for a soul. Through this ambivalence, the story presents plants as fundamentally different from and inferior to humans and animals, existing halfway between the inanimate and the animate.

Meyrink's story parallels the anatomist's perspective with the view that plant life is equivalent to human and animal life. Viewing bodies as pieces belongs especially to the perspective of the anatomist, whose work, dissecting, cutting into and taking apart resembles analyzing. The wounds on the dead man's body in the laboratory are described using the vocabulary of anatomy: "small knife cuts on chest and thighs show that he had been dissected [vivisected]" ("Kleine Messerschnitte an Brust und Schläfen zeigten, daß er seziert worden war," 85). The body parts have been, "pieced together" ("zusammengestückelt," 83), and with an, "incomprehensible art spliced together" ("unbegreiflicher Kunst zusammengefügt," 84). The way in which Dr. Cinderella reduces humans and animals to body parts is a criticism of anatomy for perceiving human and animal life as mere substance that can be taken apart and put back together.[194]

---

194 Meyrink's critique of the anatomist bears some resemblance to contemporary philosopher, Henri Bergson's views on intelligence and intuition from his book *Creative Evolution* (1911). Briefly and by no means comprehensively, the anatomist in Meyrink's story perceives the world analytically and quantititively, yet this view of life cannot access the qualititative aspect—the vital impulse, what Bergson calls the elan vital. It could be argued that the soul or spirit in Meyrink's story corresponds to Bergson's concept of intuition. Instead of awaking intuition, Dr. Cinderella puts his to sleep when he imitates the statue in an occult rite. Meyrink's association of plants with the sleeping state of Dr. Cinderella also shares some affinities with Bergson's view of the consciousness of plants in his book Creative Evolution. While Bergson asserts that there is no definite distinction between plants and animals, he does argue that movement and consciousness

This criticism of the anatomical sciences places "Die Pflanzen des Doktor Cinderella" within a collection of texts by Meyrink that criticize medicine and science. As Mohammad Qasim argues, several of Meyrink's main characters are doctors,[195] who tend to either the comical or the uncanny. Qasim draws the conclusion that these figures form a critique of the sciences through satire and grotesque imagery, when they highlight the high-handedness of scientists and the latent ability of the modern sciences to manipulate people, producing neither machines nor humans, but rather monstrous constructions of both human and machine (84). In "Die Pflanzen des Doktor Cinderella," the critique of sciences takes a slightly different perspective, divorcing human consciousness from the body, reducing it in the process to mere material. The plants, however, do not resemble machines, but rather lesser organisms incapable of possessing a soul—unnaturally created out of pieces from human and animal body pieces that once possessed a soul.

The disassociation of the narrator from Dr. Cinderella and his carnivorous plants points to a broader critique of materialism that underlies his critique of sciences. Meyrink was openly critical of materialism and its view of living beings as mechanical in nature. In his essay on yoga, "Die Verwandlung des Blutes" ("The Transformation of Blood"), he writes:

> The pure spiritual perspective will have captured a victory, when the person can harden himself and others, to the idea that there exists not just the material world in and of itself. As the Vedanta and other similar systems of knowledge teach, this is a deception of the senses—to the apparent running away with the idea of materiality.[196]

Meyrink sees materialism as a "deception of the senses," believing in the existence of a spiritual world in addition to the one apparent to the senses. The title of his essay refers to his further belief that the spiritual world can influence the physical and transform it. The complicated blend of the occult and

---

sleep in the plant as recollections, when a plant regains mobility, it also regains an equivalent consciousness (117, 124).

195 In addition to Dr Cinderella's Plants, Qasim refers to "Das Präparat" ("The Preparation") and "Das Wachsfigurenkabinett" ("Waxworks").

196 "Der Sieg der rein geistigen Anschauung wird erst errungen sein, wenn der Mensch sich selbst und andern gegenüber praktisch erhärten kann, dass Materie an sich überhaupt nicht existiert, sondern, wie der Vedanta und andere ähnliche Erkenntnissysteme lehren, eine Täuschung der Sinne bedeutet—zu scheinbarer Gegenständlichkeit geronnene Idee ist."

science in "Die Pflanzen des Doktor Cinderella" simultaneously defends the existence and potential threat of the spiritual world while criticizing materialism through the figure of the anatomist.

Meyrink's critique of science draws on the Romantic tradition of the mad scientist, whose experiments with life often reflect anxieties surrounding the medical sciences in periods of heightened anti-rationalist sentiment. Dr. Cinderella bears a special resemblance to two famous characters from 19th century Romantic fiction, Dr. Frankenstein and Dr. Moreau,[197] who also dissect bodies in order to assemble living creatures from their parts. Dr. Frankenstein composes a man out of parts from dead bodies, while Dr. Moreau forms hybrids from living animals and people. According to Rosylnn Haynes' typology of mad scientists, all two doctors are "victims of their own discovery," as is common to literature during periods of heightened anti-rationalist times (252). Both Dr. Frankenstein and Dr. Moreau die violently at the hands of their creations. Although the repercussions for Dr. Cinderella are less fatal, the laming of half his body as well as his mental fracturing carry symbolically the weight of what he has done. In Haynes' view, the predominance of these figures at the end of the 19th century suggests anxieties surrounding the manner in which science gained knowledge about the body as well as specific suspicions of the medical practitioner as a butcher. These stories also tap into the grisly history of the anatomical sciences, haunting graveyards and harvesting bodies for study.[198] The characteristics that Dr. Cinderella shares

197 Mary *Shelley's Frankenstein* or *The Modern Prometheus* (1818) and H. G. Wells' *The Island of Dr Moreau* (1896).

198 During the 18th and 19th centuries, bodies were particularly difficult to obtain for study in Great Britain and the United States. Only murderers who had been executed could be dissected in Great Britain. With few other options at their disposal, anatomists resorted to grave robbing and other illegal means of obtaining bodies to support their studies. The British Warburton Anatomy Act of 1832 was passed in reaction to the prevalence of body snatching, and allowed unclaimed bodies to be dissected for purpose of study, a practice already allowed in most European countries. At the *fin de siècle*, Vienna was one of the main centres for the anatomical sciences, their prestige supported by a steady stream of bodies from those left unclaimed by the poor. A public debate in 1903 challenged the dissection practices as a "defilement of the corpse" and "turned students into inhumane doctors," which masked largely anti-semitic political aims as respect for the dead (Buklijas 14). For a thorough discussion of anatomy in Vienna, see Tatjana Buklijas, "Cultures of Death and Politics of Corpse Supply: Anatomy in Vienna, 1848–1914" (Bull Hist Med. 2008; 82(3): 570–607.) For a discussion of the history of anatomy in Great Britain, see Helen Patricia MacDonald, *Human Remains: Dissection and Its Histories* (New Haven: Yale University Press, 2005).

with other mad scientist characters places him within the tradition of Romantic science fiction and also suggests a correlation between anti-rationalist sentiment and anti-materialism in the figure of the anatomist.[199]

Dr. Cinderella's resemblance to Dr. Moreau suggests that Dr Cinderella is as much a monster as his carnivorous plants. As typical of Haynes', "mad, bad and dangerous" scientist, Dr. Moreau and Dr. Cinderella are isolated and obsessed with their research (252). As with Dr. Moreau, Dr. Cinderella's interest in the bronze statue causes him to seem inhuman. He is overcome with a "sick curiosity" ("krankhaften Neugier") and was never before so "thirsty for knowledge" ("wissensdurstig," 72). He also describes his relationship to the statue as vampiric: "An uncanny feeling often overcame me with that: I am ruminating on something poisonous—something evil that with malicious pleasure lets me loose from the path of lifelessness only to suck at me steadfastly later like a terminal illness" (73).[200] Dr. Cinderella also comes across as asocial, speaking only with the police and the imagined reader: "Do you see, there [...]" ("Siehst du, dort [...]," 72). The vampiric overtones of his obsession with the bronze statue and his isolation associate the interests of an anatomist with the blood lust of vampires.

Meyrink draws an unusual link between anatomy and the funeral practices of ancient Egypt through the bronze statue, creating a chain of vampires that proliferate throughout the story. The bronze statue found by Dr. Cinderella in Thebes is a likeness of the god Anubis also known as the Wepwawet and is usually depicted as half-man and half-jackal. In ancient Egyptian myth, Anubis is often called the embalmer, presiding over mummification and funerary rites. He also acts as a guardian for the underworld, weighing the hearts of the recently dead before they are able to cross into the other world. On jars used for human organs in tombs, Anubis is often depicted standing on a water lily alongside two other animal-headed gods and a human-headed god. During the opening-of-the-mouth ceremony, a ritual intended to vivify the corpse, priests wore headdresses with the likeness of Anubis (Doxey, np). The narrator attributes the turn his life has taken to the statue of Anubis that acts on him like a vampire: "Like links of a chain, these haunting disturbances hang together

199 For a discussion of mad scientists in Romantic literature, see Chris Baldick. "Dangerous Discoveries and Mad Scientists: Some Late-Victorian Horrors" *In Frankenstein's Shadow: Myth, Monstrosity, and Nineteenth-Century Writing*. (Oxford Clarendon Press, 1987).

200 "Ein unheimliches Gefühl überkam mir oft dabei: ich grüble an etwas Giftigem—Bösartigem, das sich mit hämischem Behagen von mir aus dem Banne der Leblosigkeit losschälen lasse, um sich wieder später wie eine unheilbare Krankheit an mir festzusaugen."

that suck out the life force, and I follow the chain back into the past, always is the starting point the same, the bronze" ("Wie Kettenglieder hängen diese gespenstischen Beunruhigungen, die mir die Lebenskraft aussaugen, zusammen, und verfolge ich die Kette zurück in die Vergangenheit, immer ist der Ausgangspunkt derselbe: die Bronze," 72). The secret of the statue turns out to be what allows it to perpetuate itself, imitation: "A secretive, automatic imitation, an unaware, restless,—the hidden driver of all being!!" ("Ein heimliches automatisches Nachahmen, ein unbewußtes, rastloses,—der verborgene Lenker aller Wesen!!," 74). The driver is further described by the narrator as a demon that demands, "that we be like him and become his likeness" ("der da will, daß wir ihm gleichseien und sein Ebenbild warden," 74). The narrator is suggesting that Dr. Cinderella's secret activities are in imitation of the Anubis' function as the embalmer who takes bodies apart, as the one who can reanimate these pieces and as the one who can perpetuate himself. The carnivorous plants, sucking blood from vials and fat from bowls, come to resemble a repetition of the statue's feeding off of Dr. Cinderella's life force (84). As with Stoker's *Dracula*, the victims become vampires in a never ending cycle.

The manner in which Dr. Cinderella imitates the statue forms a critique of any occultist practices which result in dividing the mind from the body. According to Meyrink's essay on yoga, he understands the ecstatic state of being as in effect the departure of the mind from the body: "ecstasy literally means departure!" ("Ekstase heißt wörtlich: Austritt!," 24). Meyrink warns against such practices, believing instead that the power of the soul lies precisely in the way it can positively influence the body: "The human soul lives in the body not for the sole purpose of leaving it like someone who has stumbled into a dead end, but rather to transform the material! ("Die Seele des Menschen lebt im Körper, nicht, um ihn zu verlassen, so wie einer umkehrt, dass er in eine Sackgasse geraten ist sondern um die Materie zu verwandeln!," 15). Dr. Cinderella's misguided imitation of the statue's position produces a trance-like state and demonstrates the lasting fractious effects of willingly dividing the mind and body. The first clue appears in the description of the statue's position, which Dr. Cinderella interprets as "some kind of unknown *ecstatic* state" ("irgendeinen unbekannten *ekstatischen* Zustand," 73, italics mine). The second appears in his description of what occurs when he correctly imitates the statue with his eyes closed: "And it is as if my consciousness followed it down a horrible set of stairs—two, four, eight, always leaping over more and more steps,—with that disappeared my memory of life, and the spectre of the appearance of death lay over me" ("Und als ob mein Bewußtsein ihm nach eine ungeheure Treppe hinabfiele—zwei, vier, acht, immer mehr und mehr Stufen überspringend,—so verfiel ruckweise meine Erinnerung an das Leben, und das Gespenst des

Scheintodes legt sich über mich," 75). Dr. Cinderella is describing the departure of his soul from his body as a result of an ecstatic state, which leaves his body soulless and reduced to mere materiality. The descent of his soul into nothingness foreshadows the reduction of the human and animal body parts into pure vegetative growth in the carnivorous plants. The practice of trances for the purpose of enlightenment begins to parallel the perspective of anatomical sciences—both see bodies reduced to mere materiality.

After Dr. Cinderella imitates the Anubis statue, its form reappears throughout the story creating a complex web of echoes between the occult, the plants, anatomical sciences and the police—a motif that points towards the presence of the uncanny. The first occurs when Dr. Cinderella imitates the statue, initiating the fragmenting of his mind and body: "I then put away the ticking clocks and lay myself down, repeating the position of the arms and the hands [...] Suddenly, for me it was as if there came up a reverberating noise from my inside, as if a big stone was rolling in the depths" (75).[201] The second occurs in the laboratory with the carnivorous plants when the dead man's body takes the form of the Anubis: "In the same blink of an eye, he appeared to slide two steps down onto me, stood there suddenly upright, the arms bent upwards, the hands to the scalp. Like the Egyptian hieroglyph, the same position—the same position!" (85).[202] The third occurs at the commissionar's office, when the scribe transforms into the Anubis: "There opened a door behind me, I turned myself around and there, a tall man stood with a heron beak—an Egyptian Anubis. My eyes went black before me and the Anubis bowed before the commissionar, went to him and whispered to me 'Doctor Cinderella'" (86–7).[203] And the last, when an employee's coat slides off the coat rack, its empty sleeves take the position of the Anubis' arms: "[...] and in going by, I brushed the employee's coat, hanging on the wall. It fell down slowly and stopped hanging by the arms of the coat. Its shadow on the chalk-white wall lifted the arms upwards, over the head, and I saw, as it intended without any

201 "Stellte dann die tickenden Uhren ab und legte mich nieder, die Arm und Handstellungen wiederholend [...] Plötzlich war mir, als käme ein hallendes Geräusch aus meinem Inern empor, wie wenn eine großer Stein in die Tiefe rollt."

202 "Im selben Augenblick schien er zwei Stufen herunter auf mich zuzurutschen, stand plötzlich aufrecht da, die Arme nach oben gebogen, die Hände zum Scheitel. Wie die ägyptische Hieroglyphe, dieselbe Stellung—dieselbe Stellung!"

203 "Da ging eine Tür hinter mir, ich drehte mich um, und dort stand ein langer Mensch mit einem Reiherschnabel—ein ägyptischer Anubis. Mir wurde schwarz vor den Augen, und der Anubis machte eine Verbeugung vor dem Kommissär, ging zu ihm hin und flüsterte mir zu: 'Doktor Cinderella."

help to imitate the position of the Egyptian statue (87).[204] As Freud writes in his essay on the uncanny, *Das Unheimliche* (*The Uncanny*, 1919), repetition is a function of the uncanny as a return of what is repressed—both the possibility of reviving the dead, in this case the human and animal body parts as plants, and Dr. Cinderella's deathlike appearance when imitating the statue's pose. Through the figure of the Anubis, Meyrink associates an ancient Egyptian god with anatomy, and the supernatural with science. Just as Freud traces back the etymology of the word "unheimlich" to the point when it comes also to mean its opposite "heimlich" to uncover its meaning, Meyrink is also tracing a similar historical course—following back science and the occult to where it merges and is based on a similar fallacy—the distinctiveness of the mind and body.

In each instance, the Anubis marks a gap, a blank spot, a moment of silence that reveals the flaws in the sciences' understanding of human consciousness or soul as well as the mistakes made by many occultists, who effectively act without thinking under the influence of blind belief. It is the unspeakable secret of the Chaldear and the ancient Egyptians referred to by the narrator (75). It is his silences marked by hyphens: "———What did I want to say?" ("———Was wollte ich noch sagen?," 74), the voiceless gaze of the plants, and the silenced dead man, dissected. All these point to fragments and pieces embodied by the plants and represented by Dr. Cinderella's, "two different sides of the face" ("zwei verschiedene Gesichtshälften," 88), his lame left leg, his periods of absent-mindedness as well as his split identity as both an Egyptologist and a scientist. As with another famous mad scientist, Dr. Jekyll and Mr Hyde, the fragmented self is a consequence of dividing the conscious self from the body and is symptomatic of the world view and practices of the sciences contemporary to Meyrink.

What is left when the conscious self is divided from the body? Meyrink's essay on yoga links his reoccurring criticism of materialism to his belief that that humans live in a state of fragmented consciousness, split between "metaphysical consciousness" and "day consciousness" ("Tages- und metaphysisches Bewusstsein," 24):

204 "[...] und im Vorbeigehen streifte ich den Beamtenmantel an der Wand. Der fiel langsam herunter und blieb mit den Armeln hängen. Sein Schatten an der kalkweißen Mauer hob die Arme nach oben über den Kopf, und ich sah, wie er unbeholfen die Stellung der ägyptischen Statuette nachahmen wollte."

> [T]he inner hidden, divided from us, stranger to us in day consciousness, primeval stranger, the silenced one, stands upright in us; he is the spinal cord—the Susumna—, that in truth is meant in yoga. The outer person is divided from him, because he stands askew—somehow in a sense 'askew' to him! (35)[205]

For Meyrink, materialism is the belief that the metaphysical consciousness does not exist and all that does exist is what is perceived through the senses while conscious. As he states further on in his essay, development of the mind comes from broadening the scope of consciousness rather than diminishing, interrupting or displacing it to that which can be immediately perceived (43). When the conscious self is divided from the body, all that remains is the metaphysical consciousness.

In Meyrink's story, the metaphysical consciousness as one half of a fragmented consciousness reflects a paradigm shift at the turn of the century in the way personality is conceived, from a coherent whole to split into many drives. Emblematic of this shift is Freud's division of the personality into the ego (das Ich), the id (das Es), and the superego (das Über-Ich). Meyrink's description of the silenced figure resonates with Freud's concept of the id (das Es) as seething underneath the surface of our awareness:

> It is the dark, inaccessible part of our personality; what little we know of it we have learnt from our study of the dream-work and of the construction of neurotic symptoms, and most of that is of a negative character and can be described only as a contrast to the ego. We approach the id with analogies: we call it a chaos, a cauldron full of seething excitations. We picture it as being open at its end to somatic influences, and as there taking up into itself instinctual needs which find their psychical expression in it, but we cannot say in what substratum. (72)[206]

---

205 "der innere verborgene, von uns abgetrennte, im Tagesbewusstsein uns fremde, urfremde (!) Mensch, der Vermummte, steht gewissermaßen senkrecht in uns; er ist das Rückenmark—die Susumna—, die in Wahrheit gemeint ist im Yoga. Der äußere Mensch ist von ihm getrennt, weil er schief steht—irgendwie in einem Sinne »schief« zu ihm!"

206 "Es ist der dunkle, unzugängliche Teil unserer Persönlichkeit; das wenige, was wir von ihm wissen, haben wir durch das Studium der Traumarbeit und der neurotischen Symptombildung erfahren und das meiste davon hat negativen Charakter, läßt sich nur als Gegensatz zum Ich beschreiben. Wir nähern uns dem Es mit Vergleichen, nennen es ein Chaos, einen Kessel voll brodelnder Erregungen. Wir stellen uns vor, es sei am Ende gegen das Somatische offen, nehme da die Triebbedürfnisse in sich auf, die in ihm ihren psychischen Ausdruck finden, wir können aber nicht sagen, in welchem Substrat."

The results of ignoring or misunderstanding the metaphysical consciousness are laid explicit in "Die Pflanzen des Doktor Cinderella" through the monstrous carnivorous plants. The plants are an expression of viewing humans and animals as merely substance, in the process losing what gives them their "humanity." What remains is simply a fragment, the primeval, vegetal consciousness. The horror of the plants comes in the confrontation of the other, vegetal, consciousness within the self.

Dr. Cinderella's grotesque plants are an answer to the question, what it means to be human, by imagining life without a soul—that which gives a person or animal their inner coherence. The story lays an immense importance on recognizing that a spiritual world governed by hidden forces exists beyond conscious perception, while maintaining a note of scepticism. The plants that Dr. Cinderella discovers are after all monsters, produced while Dr. Cinderella was in a deep trance-like state of a vegetable. Through the destructive consequences of Dr. Cinderella's imitation of the statue, Meyrink equates the dangers of misguided occult practices with the materialist perspective of anatomy, showing how both lead to the division of body and mind. His carnivorous plants also subvert a hierarchical social order that distinguishes humans and animals from plants. In plant-form, human and animal body parts become living fragments of a whole, incomplete and soulless—reduced to mere vegetative growth. Dr. Cinderella follows a similar path under the statue's influence, reduced to a moving, soulless being that echoes the blurring of categories of being. Through the relationship of Dr. Cinderella to his carnivorous plants, Meyrink's short story illustrates the fear that accompanies an encounter with the familiar, and the strange, leading to the subsequent existential uncertainty.

CHAPTER 5

The Plant Bites! Deviant Plants in *Nosferatu* and *Alraune* as Metaphors for Social Instability in Weimar Culture

As is well known, the preoccupation with the occult in fantastic literature would find a continuation in the Weimar era in the realm of expressionist film. Beginning with Lotte Eisner's foundational study *The Haunted Screen* (first published in French as *L'Ecran Démonique* in 1952), scholars of Weimar cinema have frequently traced a genealogy of occult motifs leading from German romanticism through fin-de-siècle fantastic literature to the representations of occultist phenomena in film such as *The Cabinet of Dr. Caligari* (1919), *Waxworks* (1924) and *Metropolis* (1927). Given my own genealogy of "plant life" from German romanticism to the turn-of-the-century, it should come as no surprise if—alongside scientific representations of plant movement in films such as *Blumenwunder*—animated plants frequently show up as occultist motifs in the fiction films of the Weimar era. But if Weimar cinema inherited the concern for the occult from romanticism and fantastic literature, the prevalence of occult motifs in these films also registered a new experience of social anxiety. At least since Siegfried Kracauer's influential study, *From Caligari to Hitler. A Psychological History of the German Film* (1947), scholars have sought to understand the predilection for uncanny phenomena in German film of the 1920s symptomatically, i.e. as the traces of social anxieties linked to the experience of profoundly unstable social and political order after WWI. While more recent scholars might not share Kracauer's teleological interpretation of these films as premonitions of Nazism, the bulk of Weimar scholarship has adopted a framework linking these films to the experience of instability, while extending Kracauer's analysis to specific forms of social upheaval including the experience of the war and revolution,[207] economic instability and the transformations of class and gender relations.[208]

207 With the intention to "reverse the perspective of Siegfried Kracauer's influential book," Anton Kaes argues convincingly that the lasting impact of WWI as a pervasive trauma can be read even in Weimar films that don't directly address WWI (4,5). Kaes has also included Nosferatu in this category of shell shock cinema as a narrative of separation, mass death and returning home.

208 Richard McCormick's *Gender and Sexuality in Weimar Modernity* reads the style of New Objectivity as symptomatic of a reaction against instability in gender roles.

 | DOI 10.1163/9789004327177_007

In the present chapter, I want to propose a similar reading of representations of plant life in Weimar cinema. Whereas an educational-scientific film like *Das Blumenwunder* could represent plant movement as a utopian discovery (in line with Balázs' vision of film as providing an alternative to rationalist thinking), other representations of plant life in Weimar film took on a much more demonic tone. Focusing my analysis on F. W. Murnau's *Nosferatu* (1922) and Henrik Galeen's *Alraune* (1928), I argue that such "demonic" and transgressive plants in Weimar film served as metaphors for the perception of social destabilization and the loss of traditional boundaries. Moreover, by comparing two films from the beginning and end of the decade, this chapter also argues that the relation to the experience of social instability changed: whereas *Nosferatu* uses demonic plants to "naturalize" social instability, essentially showing disorder to be an unavoidable state of things, *Alraune* represents the demonic plant, which by transgressing the stable order, must be punished.

Plants appear briefly in *Nosferatu* and *Alraune*, playing small but crucial roles in films that reflect on the shifting gender and class relationships as well as the relationship of the scientist to nature. In *Nosferatu*, the protagonist Ellen empathizes with a bouquet of flowers, and the scientist, Professor Bulwer, projects a clip of a Venus flytrap devouring a fly. In *Alraune*, plants assume a more prominent place in the narrative through the title character, who is created from a mandrake root, even if the mandrake itself appears only a few times. But although plants may be on screen for only a brief time, they nonetheless serve as central metaphors for the profound social instability during the Weimar Republic, reflective of the immediate post WWI period for *Nosferatu* and the height of the stabilization period for *Alraune*, and they serve to dislocate the position and role of the monstrous.

The two films have received unequal attention by scholars with the lion's share going to *Nosferatu*.[209] Many influential scholars have analysed and interpreted *Nosferatu* within the framework of Weimar Expressionist film aesthetics, resulting in a myriad of competing readings. But despite the variety of interpretations, most readings do fall within the frameworks of expressionist film research outlined above; whereas Tom Gunning positions *Nosferatu*

209 *Alraune* has received little attention from scholars although it was immensely popular at the time (the original film with Brigitte Helm was followed by a remake in sound). A look at Valerie Weinstein's book chapter, one of the few interpretations of *Alraune*, will be included after the discussion of *Nosferatu*.

within a long history of romantic and occultist phenomena,[210] other scholars have followed Kracauer's lead,[211] linking the film to contemporary socio-political instabilities including the trauma of WWI, anti-Semitic tendencies[212] and transformations in gender roles.[213] *Nosferatu* is well known to many film scholars in German Studies for its characteristic Expressionist aesthetic and as a German adaptation of Bram Stoker's vampire novel, *Dracula*. The film closely follows the plot of Stoker's novel with a few name changes: the film begins in Wisborg with a scene of apparent domestic tranquillity between Hutter (Jonathan) and Ellen (Mina), when Hutter gives Ellen a bouquet of cut flowers. Hutter is sent to Transylvania by his employer, Knock (Renfield), the real-estate agent and assistant to Nosferatu, the vampire, to close a sale. While there, Hutter discovers Nosferatu's true nature and fears the worst when Nosferatu leaves for Wisborg. During Nosferatu's journey, Professor Bulwer (Van Helsing) is shown teaching his students about "vampirism" in nature, a lesson accompanied by clips of a Venus fly trap devouring a fly and microscopic images of menacing polyps. Nosferatu proceeds to terrorize the town until Ellen, educated by vampire lore, offers herself as an ecstatic sacrifice in order to kill Nosferatu by luring him into the daylight.

Alraune has also been adapted from a book of the same name by Hanns Heinz Ewers, although with considerable changes. The film begins with a scene depicting the collection of the mandrake root from underneath the gallows at midnight, shifting dramatically to a scene between Professor Jakob ten Brinken and a group of scientists as he announces his plan to genetically engineer a human being based on the legend of the mandrake root and with

---

210 In his essay "To Scan a Ghost," Gunning reflects in particular on the relationship between the tradition of the uncanny and the proliferation of ghosts and phantoms in "new" visual and auditory media.

211 In *From Caligari to Hitler*, Kracauer saw in Nosferatu the looming threat of tyranny that could only be staved off through love (77–78).

212 John Sandford follows up on Kracauer's thesis, reading a deep anxiety about external chaos disrupting German society. Sandfords draws parallels between the association of *Nosferatu* with rats and the anti-semitic film *Der Ewige Jude* (1940) to demonstrate the roots of the NSDAP period in the Weimar Republic. Kaes also interprets *Nosferatu* from this angle in his contribution on Weimar film in *Geschichte des deutschen Films* (52).

213 Both Janet Bergstrom and Judith Mayne discuss *Nosferatu* and Weimar film from the perspective of gender. Bergstrom argues that class distinctions and typing was as important to Weimar film as gender distinctions (189), yet in Murnau's films and specifically *Nosferatu* the erotics of looking is displaced from solely the female body to include the male body and even landscape.

the help of a criminal's semen and a prostitute. The result, Alraune, is raised by Brinken, educated by nuns and rebels by running off with a male friend, whom she convinces to take money from his father's workplace at a bank. On the train, they meet and join a travelling circus. The circus master becomes Alraune's next conquest, soon to be left behind as she discovers the lion tamer and subsequently shows him how to tame the wild cats. Alraune is rediscovered by her "father," performing as a disappearing bird in a magician's act, and declared to be rescued by him from infamy. They proceed to travel from town to town accumulating expenses until Alraune discovers the origins of her birth and begins to plan her escape. Brinken's final act of desperation, an attempt to kill Alraune, ultimately drives her into the arms of an aristocratic viscount.

During the seven-year gap separating *Nosferatu* (1921) and *Alraune* (1928), a shift occurred in Weimar film from an emphasis on Expressionist film to a more realist aesthetic, New Objectivity (Neue Sachlichkeit). This stylistic shift to New Objectivity, and the greater stability in the Weimar Republic after the introduction of the Dawes Plan, correlate to a change in the visibility of social instability.[214] Richard McCormick in his authoritative book, *Gender and Sexuality in Weimar Modernity*, sees in New Objectivity "an obvious gesture of disavowal of the underlying anxieties about gender and modernity, an attempt to re-achieve 'masculine' mastery through objectivity, science, technology. [...] Mastery would be regained by documenting the anxieties of modernity 'objectively' and 'soberly' with the help of modern technology and/or 'scientific' methods" (51). There was also an overwhelming concern with "surface reality" and "a disavowal of 'inwardness'" (53). In comparison to *Nosferatu*, the subversive elements of plants in *Alraune* appear in the breaks of the New Objectivity aesthetic, reflecting the suppression of latent feelings of instability rather than overt instability. In contrast, the use of Expressionism in *Nosferatu* corresponds to the visibility of social instability and fluidity.

The relationship of plants to social instability in *Nosferatu* and *Alraune* is best understood when measured against one of the most persistent ways humans organize nature, the scala naturae. Also known as the great chain of being, the scala naturae organizes organic life and inorganic matter in a linear hierarchy from the inanimate material, rocks and crystals, to plants, animals, humans and finally the divine. The concept of organizing nature from lesser forms to higher forms in a linear hierarchy has existed in some form or extent since ancient Greece.[215] Aristotle's division of the soul in three degrees

214 McCormick notes that the rise of New Objectivity is commonly associated with the greater stabilization period in the Weimar Republic from 1924 to as late as 1933.

215 Arthur Lovejoy's *The Great Chain of Being* from 1936 was and still is one of the most influential histories of the idea.

from nutritive (plant), perceptive (animal), and rational (humans) is a formative example of a linear hierarchy. In a famous essay from the early 20th century, Emile Durkheim and Marcel Mauss suggest that the way people organize nature reflects to some extent the way they organize social relations.[216] In literary studies and history, the chain of being has been often used as a model for interpreting hierarchical relationships.[217] Applying this model to *Nosferatu* and *Alraune* reveals the ways in which disruptions to the hierarchy that locates plants below animals and humans reflects the sense of social instability in the Weimar Republic.

The concept of the scala naturae was present in the Weimar Republic, albeit greatly impacted by Charles Darwin's theory of evolution and discoveries in the fossil record. With Darwin's theory of evolution, the understanding of life forms changed from a model of stable being to one of becoming (and going extinct), giving the great chain of being a sense of increased mobility and adding the idea of progression. The impact of evolution can be seen in Ernst Haeckel's famous "biogenetic law" (which stipulated that individual life forms pass through their evolutionary ancestry) at the turn of the century.[218] Representations of the chain of being were also common in Weimar educational films. The culture film, *Nature and Love* (*Natur und Liebe*, 1927), for example, traces life from its most primitive forms such as the polyp through

---

216 Mauss and Durkheim's essay "Primitive Classification" was originally published in French in 1903 and argued that nature was not based on an a priori logical hierarchy, but rather on the social hierarchy: "According to him [Frazer], men were divided into clans by a pre-existing classification of things; but, quite on the contrary, they classified things because they were divided into clans" (127).

217 Both Alice Kuzniar's article "A Higher Language: Novalis on Communion with Animals" and Christopher Clason's book chapter "Automatons and Animals: Romantically Manipulating the Chain of Being in E. T. A. Hoffmann's 'Der Sandmann' and *Kater Curr*" demonstrate how the chain of being is used as a model for interpreting literature.

The chain of being continues to shape the way we perceive nature. In an article for *Nature* magazine, Sean Nee analyzes recent texts by Richard Dawkins (2004), John Maynard Smith, and Eors Szathmáry (1995) among others, and finds that the concept of the chain of being is still present as a means of organizing nature.

218 In his entry for the *Encyclopedia of Race and Racism*, Jonathan Marks summarizes Ernst Haeckel's variation of the chain of being as follows: "Darwin's German apostle, Ernst Haeckel, would go further, constructing a theory of evolution that stretched from the amoeba to the German nation, driven by his 'biogenetic law' (that ontogeny recapitulates phylogeny, or that individuals personally pass through developmental stages representing their ancestry). In such a grand view, not only would other races be primitive and inferior, but so would other social institutions and political systems" (72).

progressively more complex beings until it reaches humans.[219] Humans are further divided into the earliest cave dwelling ancestors through the Iron Age up until the final futuristic image, a projection of the ultimate progression through the visual metaphor of a ladder, leading into the light. The last image of people climbing the ladder into the light makes explicit the assumptions that evolution leads to progress and that humans are at the top of a linear hierarchy as the highest forms below the "light." Films like *Nature and Love* expose a cultural narrative informing early 20th century representations of nature, which organizes the natural world according to an anthropocentric perspective that values progress and rationality (symbolized by the light). *Nosferatu* and *Alraune*—and more centrally their plant representations—present the underside to the story of progress. Through representations of dangerous plants, the films thematize both the fear induced by changing social relations and the possibility of devolution or degeneracy.

As has been observed by many feminist theorists, the chain of being further divides humans into hierarchical subsections with men above women. In the opening sequence of *Nosferatu*, this gender hierarchy is set up only for ambiguity to be introduced at the end of the scene. The establishing shot of the film, the tower in Wisborg (Figure 23), is immediately followed by a shot of the film's main protagonist, Hutter, grooming himself before a mirror (Figure 24). Through this associative montage, the film connects the two images thematically and graphically. The tower is significant for its cultural connotations as a symbol for the achievement of human civilization and progress but also for language and power. Matched by the subsequent shot of Hutter, the tower's connotations become associated with Hutter and consequently masculinity.

In contrast with Hutter's proximity to the tower, the subsequent sequence associates Ellen more closely with nature than the city and positions her as the object of Hutter's gaze. After Hutter is finished grooming, he goes to look out of the window. The following eye-line matching shot of Ellen looking out the window surprises the viewer, who expects to see a shot of the street or of the city. Instead, they are faced with a shot that directs outward, but is actually looking inward. Looking into Ellen's space we see an image of domesticated nature, flowers as decorations in window boxes and a cat whose purpose, it seems, is to be Ellen's plaything (Figure 25). The prominence of plants frame

219 The film also has four possible subtitles which reveal a great deal about the film's content and narrative arc: "Ein Film von Liebe und lebendigem Werden" (A Film about Love and Living Becoming), "Schöpferin Natur" (Creator Nature), "Vom Urtier zum Menschen" (From Primeval Animal to Human), and "Von der Urzelle zum Menschen" (From Primeval Cell to Human).

FIGURE 23 *Tower in Wisborg from* Nosferatu. *Scene still. Dir. F. W. Murnau. Minneapolis, Minn.: Mill Creek Entertainment, LLC, 2006. Illustration reproduction courtesy of the* Friedrich-Wilhelm-Murnau-Stiftung.

Ellen and mask the window frame, creating, for a brief moment, ambivalence as to whether she is outside—in the countryside or inside a building in town. In contrast, Hutter's bare walls and simple window dressings unequivocally place him inside a building and within a town. The ambivalence of Ellen's location in addition to the prominence of plants in this first image of Ellen places her closer to nature than Hutter. In the subsequent shot, the floral print on the wallpaper and the framed picture of plants visible in the background further reinforces Ellen's closer connection to nature. This brief sequence and spatial framework embedded in these opening images suggests a hierarchical division between men and women that conveys traditional, conservative values.

The stability of social relations at the beginning of the film is reflected in the typical depiction of the chain of being as seen in the behaviour of the cat and the plants surrounding Ellen as well as the composition in the frame. Central within the frame, Ellen is higher up on the ladder than the cats and plants. From her interaction with the cat, it is clear that she is in control, manipulating the cat's movements with the use of a pendulum. Like the cat, the plants are defined in terms of their relationship to her. They serve as a visual frame

FIGURE 24 *Hutter Grooming from* Nosferatu. *Scene still. Dir. F. W. Murnau. Minneapolis, Minn.: Mill Creek Entertainment, LLC, 2006. Illustration reproduction courtesy of the* Friedrich-Wilhelm-Murnau-Stiftung.

and as decoration. The plants are static, mere ornamentation, resembling more inanimate objects than living beings. Passive and with no sign of self-initiated movement, the plants in the window box differ in no great way from the plants visible on the wallpaper and in the framed picture behind Ellen. The plants here are enframed, contained, unlike the plants later in the film. The perception of plants as discrete, stable objects that have been instrumentalized is set up in this scene only to be later destabilized by an alternative perception of nature as fluid and possessing menacing agency. In addition to the decorative use of plants, the use of plants as a form of symbolic currency becomes apparent in the next sequence when Hutter goes outside to cut some flowers for Ellen. He is shown cutting flowers, surrounded by bushes that appear wild, inviting a contrast between his civilizing influence and nature and foreshadowing the conflict of interests between the social use of plants and the perception of plants as living beings. Hutter's attempt to express his love for Ellen is not unusual behaviour, but represents a widespread instrumentalization of flowers. A short film from the 1920s, *Laßt Blumen Sprechen* (*Let Flowers*

FIGURE 25 *Ellen at the Window from* Nosferatu. *Scene still. Dir. F. W. Murnau. Minneapolis, Minn.: Mill Creek Entertainment,* LLC, *2006. Illustration reproduction courtesy of the* Friedrich-Wilhelm-Murnau-Stiftung.

Speak),[220] playfully taps into the popularity of giving flowers and refers to the plethora of books published during the 19th century discussing the language of flowers.[221] The short film shows two men competing for a woman's affections by giving her increasingly larger bouquets of flowers. In *Nosferatu*, the joy apparent on Hutter's face while cutting the flowers reveals his expectation that Ellen will participate willingly in this exchange of symbolic currency and reinforce his perspective that flowers exist for the purpose of human use (Figure 26).

Instead, Ellen's reaction to Hutter's flower bouquet subverts this instrumental use of flowers and introduces one of the central themes of the film: her reaction questions the traditional gender hierarchy through introducing

220 The film was released in Germany in 1929 according to the censorship cards, but appeared in France some time earlier under the title: *Quand Fleurs Parlent.*

221 The Walter Ruttmann animation, *Das wiedergefundene Paradies* (*Paradise Regained*), from 1925 is another example of a flower language used as an advertisement.

FIGURE 26 *Hutter Gives Ellen Flowers from* Nosferatu. *Scene still. Dir. F. W. Murnau. Minneapolis, Minn.: Mill Creek Entertainment, LLC, 2006. Illustration reproduction courtesy of the* Friedrich-Wilhelm-Murnau-Stiftung.

instability into the hierarchy of being. Ellen responds to Hutter's gift of the flower bouquet as if the flowers were human instead of reading his gift as a sign of affection. She turns away from Hutter, placing herself between him, the perpetrator, and the flowers, the victims. Then she cradles the flowers, caressing them as if they were a baby (Figure 27). Her empathy with the cut-down flowers is made explicit in the following intertitle, when she asks Hutter, "Why did you kill them, the lovely flowers?" By elevating the flowers to the status of a living being that can be killed, Ellen reveals the horror in the common use of plants as tokens of affection and thereby suggests that the scala naturae is itself founded upon violence. She also distinguishes her view of nature as fundamentally different from Hutter's, whose naivete contrasts with her greater awareness.[222] As we saw in chapter two, Ellen's reaction to the flowers is repeated in the later film, *Das Blumenwunder* (*The Miracle of Flowers*), when

222 I am referring to the scene when she sits up suddenly in bed, reaches her arm and calls for Hutter, suggesting that she is aware of what is occurring a great distance away.

FIGURE 27 *Ellen Cradling the Flowers from* Nosferatu. *Scene still. Dir. F. W. Murnau. Minneapolis, Minn.: Mill Creek Entertainment,* LLC, *2006. Illustration reproduction courtesy of the* Friedrich-Wilhelm-Murnau-Stiftung.

Flora, the fairy, stops the children from destroying a flower bed and equates the life of flowers with that of humans: "The flowers have life like yours" ("Die Blumen haben Leben gleich Euch"). As with Flora's reaction, Ellen's also reverses the perspective that views plants as inanimate objects, thus questioning the objectification of the plants and ultimately the objectification of herself by Hutter. As the film progresses, the distinctions between plant, animal and human will become increasingly blurred, challenging the hierarchy of man over woman and humans over nature.

The use of plants as a metaphor for instability in gender roles conforms to a broader subversion of the standard relations between active men and passive women in the film that has been identified by scholars such as Janet Bergstrom. Instead of Ellen as the sole, passive object of desire, the film introduces fluidity into the relationship between desiring and being desired, what Bergstrom has termed an "erotics of looking." She reads the reclining pose of Hutter, after his first night in the castle as an example of the male body as passive, feminine, and momentarily powerless without undermining his heterosexuality (197). She connects this scene to a diffusion of visual pleasure in *Nosferatu* that includes women, men and even landscapes. By focusing on this diffusion of visual pleasure, Bergstrom's concept of the "erotics of looking" reveals the fluidity and ambiguity present in the relationship between Ellen, Hutter, Nosferatu and the spectator and in their shifting roles.

A further layer of complexity is added to the shifting and ambiguous social hierarchy in a famous montage sequence in which Professor Bulwer gives a scientific lecture on vampire-like phenomena in nature. Through the three hybrid figures—the Venus flytrap, Knock (the real estate agent) and a polyp—the film codes fluidity of social status and, consequently, deviance and degeneracy as natural. The montage sequence is set up as an example of the laws of nature. In an introduction to Professor Bulwer, the purpose of his investigations into nature is identified—in a phrase with Haeckelian overtones—as finding the "unifying principle of nature." The coding of the images as natural is continued in the introduction to the Venus flytrap. We learn from the narrator in an intertitle that Professor Bulwer is teaching his students about the "cruel habits of carnivorous plants." The intertitle continues: "In horror, the students observe Nature's mysterious ways." Bulwer's lesson occurs as Nosferatu and his soil-filled coffins are being transported to Wisborg on the ship named Empusa.[223] At the same time, the shots are also intercut with images of Knock who begins to imitate the spiders he sees in his cell by eating insects and crying out "blood is life!" The paralleling action positions the Venus flytrap as a metaphor for Nosferatu and suggests that his "cruel habits" and "mysterious ways" are as much a part of nature—as representative of nature's "unifying principle"—as are carnivorous plants and beastly humans such as Knock.

The following shot of the Venus flytrap (Figure 28) closing in on a fly stands out for its realistic aesthetic—the footage may have been taken from a contemporary nature film (or modelled on one)—in an expressionist film with a prominent use of chiaroscuro. As with many other plant films from the 1920s, the image reflects the unemotional gaze of the scientist, whose distance to the object of study contrasts with the emotional expressivity and interiority seen throughout the film. As Tom Gunning argues, the image of the Venus flytrap is "the visualization of scientific mastery" (Gunning, "To Scan a Ghost," 98). The unmodified shot of the Venus flytrap appears to conflate science and film,

---

223 The ship's name, Empusa, is the first subtle detail that connects this sequence to hybridity. Empusa was a goddess in Ancient Greek mythology, who was known for her ability to assume many different shapes and to suck the blood of travellers passing by. This version of the myth comes from chapter four of *Life of Apollonius of Tyana* by Philostratus.

FIGURE 28 *Venus flytrap from* Nosferatu. *Scene still. Dir. F. W. Murnau. Minneapolis, Minn.: Mill Creek Entertainment, LLC, 2006. Illustration reproduction courtesy of the* Friedrich-Wilhelm-Murnau-Stiftung.

suggesting that both the scientist and the filmmaker have an unquestioning grasp of what constitutes reality. Yet Bulwer is also powerless to stop nature's cruel habits. As Judith Mayne observes, the figure of Bulwer "suggests that the forces of science, reason, and civilization can no longer successfully wage battles against the Draculas of the world, but exist only to give illustrated lectures" (30). The realistic aesthetic of the Venus flytrap exhibits the strengths of science, the ability to shift and change the way we perceive nature through documenting the ways in which the natural world thwarts expectations such as the existence of a plant that eats flies instead of being eaten. Yet the much diminished role of Bulwer compared to Bram Stoker's scientist Van Helsing also illustrates the limits of science, failing to address the affective response to shifting societal norms. In response to the "unplant-like" behaviour of the Venus flytrap, the students observe "in horror".

The students' horrified response to the Venus flytrap is part of larger cultural narrative surrounding carnivorous plants that labels their behaviour as deviant and degenerate. The surprising and sensationalist behaviour of carnivorous plants inspired many short nature films alongside other oddities from the plant world.[224] A short film also from 1921, *Die Seele der Pflanzen* (*The Soul of Plants*), compares the Venus flytrap to a vampire: "In warm, humid days, the plant becomes a vampire" ("In feuchtwarmen Tagen aber wird die Pflanze zum Vampyr"). The film continues on to elicit sympathy for the insect caught in its grasp: "Pity the victim that isn't strong enough to escape the closing bars" ("Wehe dem Opfer, das nicht stark genug ist, den schließenden Gittern zu entrinnen"). Often, neutral language is mixed with animal metaphors and comparisons to criminality. In a short film from 1925, *Wunder der Pflanzenwelt* (*Wonders of the Plant World*), a fly caught in a Venus flytrap is referred to as "prey" ("Beute"), which can also be translated as booty or loot. The connection to criminality is more explicit from the title of a film from 1930, called *Räuber in der Natur.* (*Fleischfressende Pflanzen*) (*Robbers in Nature* (*Flesh-Eating Plants*)). In the same film, the trapping motion of Venus flytrap is described through an animal metaphor: "A fat bite" ("Ein fetter Bissen"), recalling Professor Bulwer's comparison to vampires after the shot of the Venus flytrap in *Nosferatu*. The way carnivorous plants are contextualized in the short plant films and *Nosferatu* reveal a perception of these plants as socially deviant, and as "criminals" because they behave like animals. These associations with criminality and social deviance characterize the plants perceived upward mobility on the chain of being as antisocial. A similar condemnation is applied to Knock's animal-like behaviour as an example of degeneracy, first, when he is interned in an asylum and, later, when he is killed by a town mob.

Bulwer's second projection, the polyp (Figure 29), presents a second example of species ambiguity. As primitive organisms possessing a simple

224 The film, *Wonders of the Plant World*, exemplifies the choice of unusual plants. The film features the *Mimosa pudica*, a plant that moves upon contact; carnivorous plants; the Victoria regia, a lily pad with unusual strength and quick growth; the Lotus flower, which only flowers for a day; the giant Redwood tree with its quick growth and extreme heights; among others.

For most of the carnivorous plant films, only the censorship cards were available for viewing. The films are as follows: Fleishfressende Pflanzen (1922) (Flesh-eating Plants), Wunder der Pflanzenwelt (1925) (Wonders of the Plant World), and Insektenfressende Pflanzen (1929) (Insect-Eating Plants), Räuber in der Natur. (Fleischfressende Pflanzen) (1930) (Robbers in Nature (Flesh-Eating Plants), and Fleischfressende Pflanzen (1943) (Flesh-Eating plants). The film from 1943 was the only one available for viewing in the state archive in Berlin. Despite its use of colour film and sound, the shot of the Venus flytrap was very much similar to the one in Nosferatu. It can be assumed that the aesthetics of the short nature films had not changed a great deal between 1921 and 1943.

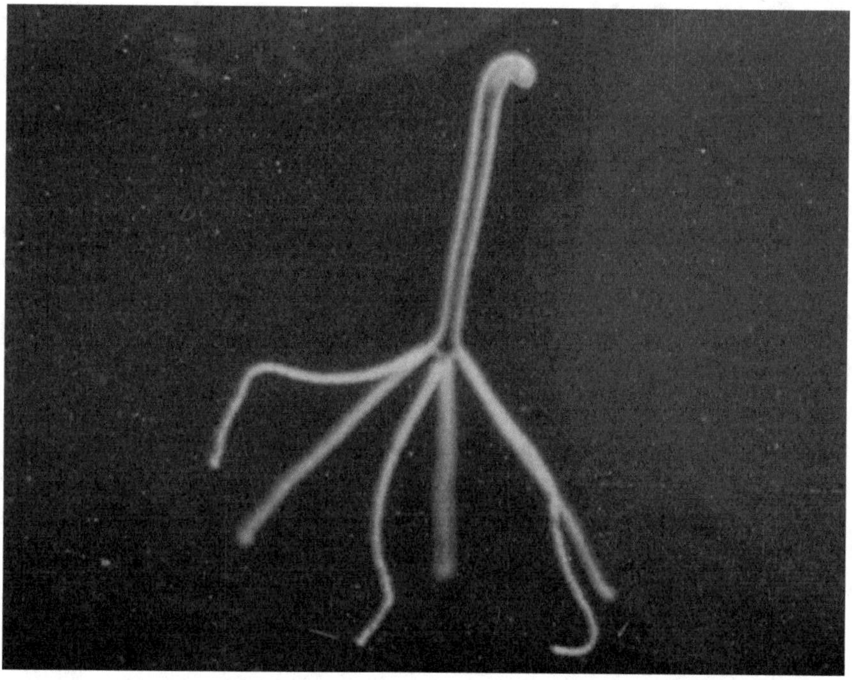

FIGURE 29 *Polyp from* Nosferatu. *Scene still. Dir. F. W. Murnau. Minneapolis, Minn.: Mill Creek Entertainment, LLC, 2006. Illustration reproduction courtesy of the* Friedrich-Wilhelm-Murnau-Stiftung.

digestion system and branchlike tentacles, polyps were often perceived as closer to plants than other more complex animals. Gustav Fechner in his discussion of the plant soul from the mid 19th century refers to polyps as "half plant-type Nature" ("halb pflanzenartiger Natur") and plants as "only wooden polyps" ("nur verholzte Polypen," 244). Polyps were also compared to plants in a short nature film from 1921, *Natur im Film: Die Hydra des Süßwassers* (*Nature in Film: The Hydra of the Fresh Water*): "The hydra almost appears like a little plant" ("Die Hydra sieht fast wie ein Pflänzchen aus").[225] As in the short films on carnivorous plants, the polyps in *Nosferatu* are referred to as "Robbers" ("Räuber"), personified as belonging the criminal class and therefore deviant. Although the polyp, the Venus flytrap and Knock are considered deviant, the film resists labelling them unnatural. Instead, deviance and destructive influences have a place within the natural system as a catalyst for social change.

225 The hydra is the genus of a polyp. The microscopic shots of the polyp in the short film *Hydra des Süßwassers* are very similar to the one in *Nosferatu*.

Viewed in this light, the final image of the film, a shot of vegetation slowly eating a stone tower, is telling for its powerful statement on the impermanence of social relations and structure.

The Venus flytrap and the polyp belong within a chain of metaphors, including Count Orlok, that reflect social instability. His animal and plant-like characteristics point to a fluidity between human, animal and plant that represent a threat to social stability and to Ellen and Hutter's relationship. The rat-like countenance of Count Orlok and his proximity to rats has been discussed by John Sandford. Building on Kracauer's thesis that the aesthetics of Weimar cinema register deep-seated anxieties about social order, Sandford argues, it is possible to detect the "germ of a more specifically racial insecurity" (i.e. anti-Semitic fears) in certain images from *Nosferatu* and other films produced during the Weimar Republic (322). Nosferatu's appearance and his origins in the east have been often cited as a reference to the Jew as a foreigner and the fear of him as a reference to the fear of a foreigner among us.[226] Kenneth Calhoon's proposes a broader reading, placing Nosferatu as the conduit between the animate and inanimate world, a place typically occupied by plants on the chain of being:

> Behind the experience of the Uncanny is a magic that undoes the border between self and world and recalls the dead to life. An insect's assimilation to a leaf or twig is evidence of a "magical tendency" in the biological world. The sickly-pale Nosferatu, who sleeps by day and eats nothing, leads the "reduced existence" of the organism that adapts itself to the inanimate. (647)

The aspect that Calhoon identifies as magical, the undoing of the border between self and world, is a metaphor for social change. The causes of social instability in the Weimar Republic were numerous, but most were linked to World War I and its aftermath. Anton Kaes has identified the film *Nosferatu* as a possible example of shell shock cinema, in which the cause for the social upheaval is the war. Common to most readings of the figure *Nosferatu* is the threat he poses to the social stability as a catalyst for chaos, whether that is from World War I trauma, racial anxieties, or a diffusion of desire. Perhaps the most radical aspect of *Nosferatu* is the framing of social upheaval and the resulting shifts in social structure as natural, causing the vampire to be seen not as the other but embedded within the social norms.

226 See Paul Monaco. *Cinema and Society in France and Germany*, 1919–1929. (New York: Elsevier, 1976).

While the transgressive Venus flytrap in *Nosferatu* frames social instability and change as natural processes, in *Alraune*, the title character's rise from the lower classes is framed as unnatural, revealing a latent resistance to social change. Professor Jakob ten Brinken's experiments with artificially breeding a woman, Alraune, are explicitly vilified in the film as going against nature. Before Brinken begins his experiments, his nephew accuses him of "violating nature" and warns that "nature will have her revenge."[227] At the conclusion of the film, the narrator repeats the judgement that he had "violated nature" and names his sentence, suffering "the hell of loneliness and insanity." The reframing of instability in *Alraune* as unnatural reveals a fundamental shift in how social change is viewed. In *Nosferatu*, the naturalization of the Venus flytrap includes their deviance from plant norms within a larger framework of unavoidable systematic change. In contrast, Brinken's experiments with artificial insemination are in fact experiments with social class that reflects a belief that upward social mobility, changing gender roles or even flux within the social hierarchy is unnatural. His punishment, diminished social status and isolation, is in accordance with his perceived crime against societal norms.

The film follows the experiments of geneticist, Professor Jakob ten Brinken with artificial insemination and parenting. He creates Alraune with help of a Mandrake root, a prostitute and the semen from a hanged man. As he states, he raises his creation without the interfering emotions from biological parentage and takes meticulous notes on her progress, kept in a journal. Alraune seems at times to exhibit some of the traits of her dubious parentage: promiscuity from her prostitute mother and devious behaviour from her criminal father. Still, as one reviewer remarks, Henrik Galeen, the writer and director, has sanitized the character Alraune in his adaptation of Hanns Heinz Ewers' novel of the same name (*Alraune*).[228] Her overt vampire tendencies have all but disappeared and, in the end, she seems to find respectability in marriage. Professor ten Brinken, on the other hand, has fallen from a position of respectability and authority to one of financial ruin, isolation and insanity.

Very little has been written specifically on *Alraune*, especially considering the star power of Brigitte Helm, who plays the title role. Valerie Weinstein in her comprehensive reading of *Alraune* argues that the film "preys on fears of racial pollution and anxieties about the New Woman and debunks science

227 All subtitles are taken from the English version of *Alraune*.

228 While there is no room here for a comparison to the novel, the reviewer's remarks can be easily justified. In Ewers' novel, Alraune is overtly vampiric, drawing blood during sex through biting and scratching, and the men around her experience not just financial ruin but often death after a relationship with her.

as an effective source of knowledge" (198). In addition, she sees the stylistic and thematic blending of horror and science in the film as corresponding to the blending of Expressionism and New Objectivity (199). Finally, Alraune is not just a product of artificial insemination but an example of "fantastic biology" that blends "biological, technological, and fantastical parentage" (204). One of the central questions of the film, according to Weinstein, is whether an experiment and observation can determine if "the parents' genetic makeup has a purely random effect on the offspring." Weinstein concludes that "the film suggests that Brinken's question cannot be answered definitively through observation" (208). Expanding on her reading, I argue that the narrative of a plant becoming human functions as a metaphor for social mobility and gender instability, which depicts both forms of social change as artificial while conversely naturalizing a rigid social hierarchy as a natural order.

From the outset of the film, a social hierarchy that subdivides people into uncivilized and civilized is readily apparent through visual clues that reveal a "naturalized" right of members of certain social class to instrumentalize others. The visualization of the Mandrake myth behind Brinken's experiment associates lower classes with nature through dark Expressionist images of a man hanging on the gallows above a rocky, wild landscape (Figure 30). According to the myth, the dead criminal's semen should be collected from the ground at midnight to create the mandrake root. It is a given that the criminal's semen could be taken disregarding any claim a criminal and a dead man might have to the use of his body. Similarly, the title that the film gives to the myth: "The Story of a Mandrake Root Which Turned into a Human Being" suggests that Alraune should be read as progressing from subhuman to human. By the end of the film, she has attained a higher degree of culture. Other appearances of people from the lower classes reiterate the early association between lower (or subhuman) social status and instrumental use. The luring of the prostitute from the street and her use as a surrogate mother illustrates a perception of women's bodies as both open for instrumental use by anyone and her subsequent invisibility (she disappears into Brinken's unseen laboratory to not be seen again) reflects the invisibility of these classes.

The question of class and status also informs the film's aesthetics. The contrast between the opening expressionistic shots and the subsequent introduction to Professor ten Brinken is striking for its abrupt shift to a realist aesthetic, New Objectivity, and for the division it suggests between the lower classes and the respectability of the professor. Realism helps to code him as well-respected member of society. He is introduced through the narrator as a "world-famous authority on genetic cross-breeding" and "his Excellency the Privy Councillor." A group of white men in suits are shown standing around him and paying

FIGURE 30 *Man Hanging Over Expressionist Landscape from* Alraune. *Scene still. Dir. Henrik Galeen. Ama-Film, 1928. Illustration reproduction courtesy of the* Friedrich-Wilhelm-Murnau-Stiftung.

deference to his explanation of his experiment with the Mandrake root. The value that Brinken ascribes to reason and science over the irrational and myth is clear: "Mandrake. daughter of the hanged man from the dark days of superstition… I shall lead you into the clear light of science." His intent is undermined by the affinity he has for the lower classes as evidenced by his caresses of the woman-shaped root and the social taboo of his incestuous feelings for Alraune. The importance of reputation and the social status to Brinken is repeated after the circus scandal: "You have misused my name. We shall travel far. Far Away. Until your past is buried."[229] Science, respectable society and the realist aesthetic are further bound together when Brinken records in his journal that he then "Introduced her into the 'right circles'" after they leave from the circus. The opening scene alongside subsequent

229 An article from 1931 in *Das Kriminal-Magazin* asserts that it is a common belief that criminality is common in the circus: "Durch diese ausgiebige Berichterstattungen steht man allgemein die Ansicht, daß kriminelle Delikte im Zirkus häufig sind" (1667).

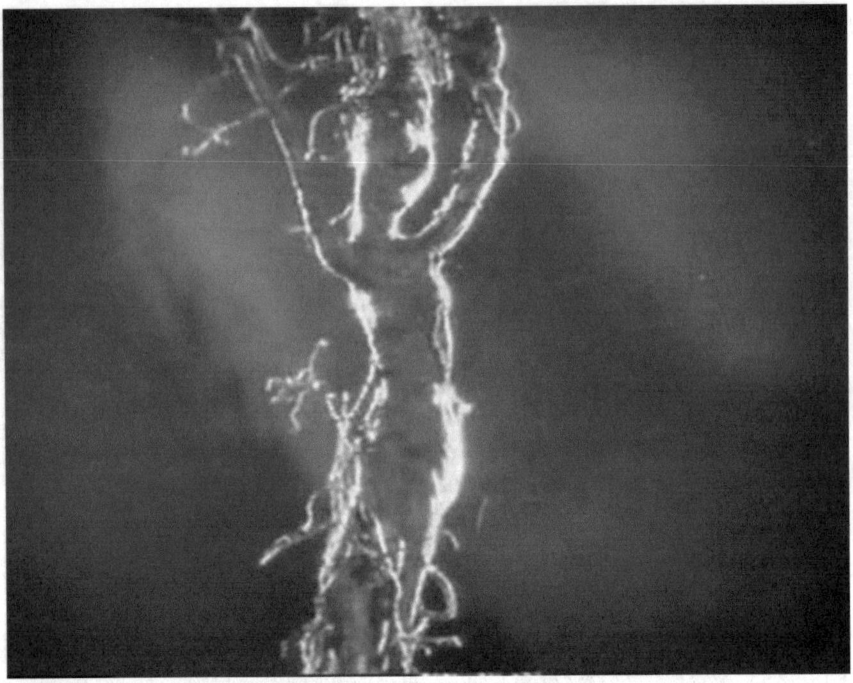

FIGURE 31 *Mandrake Root from* Alraune. *Scene still. Dir. Henrik Galeen. Ama-Film, 1928. Illustration reproduction courtesy of the* Friedrich-Wilhelm-Murnau-Stiftung.

reminders plays a crucial role in the film by setting the societal norm against which Alraune's and Brinken's subsequent behaviours can be judged.

For Alraune, the question remains throughout the film whether she has been properly socialized to become a lawful member of society and if her fathers' crimes, the criminal's and Brinken's, leave a mark on her. On the surface, she appears to have been fully assimilated: she's an attractive, athletic woman with blond hair and blue eyes, but the film frequently undermines this outward appearance via other visual clues, giving cause to read her, as does Weinstein, as the "embodiment of the mandrake root" (203). Weinstein mentions the two most overt connections: Brinken fondles the human-like mandrake root near the start of the film, and at the point of his downfall, he visualizes the root rotating as it dissolves into a woman's body (Figure 31 & 32). This dissolve, as Weinstein reads it, "blurs the boundaries between magic object and the vamp." She continues: "The film explicitly codes the mixings that created Alraune as impure" (203). The dissolve also illustrates Alraune's social mobility as a fluid transformation not necessarily controlled by Brinken but witnessed by him.

FIGURE 32 *Alraune from* Alraune. *Scene still. Dir. Henrik Galeen. Ama-Film, 1928. Illustration reproduction courtesy of the* Friedrich-Wilhelm-Murnau-Stiftung.

A poster advertising *Alraune* also visualizes her impurity as a contrast between a human ascending to the light and the roots of a plant reaching down into the darkness. In the top right corner of the poster Alraune's head is visible, but instead of a body there are roots, reaching down into the darkness of the soil and wrapping around the decapitated heads of her "victims." The deadly position of the root tendrils recalls the deviant behaviour of the grasping of the Venus flytrap and polyp in *Nosferatu* but also the fluidity of their state of being as between animal and plant. The image also gives Alraune an oddly animalistic character as her roots resemble an octopus grasping its prey, a precursor to the repeated associations between Alraune and animals throughout the film. A scene that was not included in the final version also suggests racial impurity. A black man is shown climbing out from the brothel of Alraune's mother, suggesting a further interpretation of Alraune's appearance. Both the film poster and the dissolve call into question the surface appearance of Alraune as a woman, suggesting an underlying instability to her state of being and social position.

Numerous other visual clues hint that Alraune's inner nature is in contrast with the respectability and status her appearance affords her.[230] She channels a child-like, primitive sexuality seen in her fearlessness around vermin. The first image we see of Alraune is reminiscent of the first shot of Ellen in *Nosferatu*. As with Ellen, Alraune is shown at a window and against a background of plants. Instead of Ellen's innocent play with a cat, however, Alraune is shown playing with an ant, repeatedly pushing it back into a dish filled with water. It is a small cruelty on her part, but does not necessarily reveal her to be monstrous. The same applies to her trick on the nun, when she places a large, hairy spider on the woman's habit. The sexual undertones to these animal associations become apparent when Alraune meets the circus manager in the train. She reacts with delight to a mouse running up her skirt, to the surprise of the circus manager, who expected fear or horror. Her association with animals retains an innocent quality through her child-like playfulness that resists reading these scenes as evidence of her monstrous character. Her association with these less desirable animals differs greatly from Nosferatu's association with rats and the hyena, which explicitly code him as monstrous. Nor does she humanize the animals as does Ellen the flowers. Rather, her association with vermin expresses her ambivalent relationship to the implications of her deviant origins on her performance of gender and her social class.

The association between Alraune and animals reaches a new level of complexity in the episode with the circus that visually codes Alraune's place in society as ambivalent and, therefore, problematic. In addition to her association with a caged bird, Alraune is also graphically and thematically compared to a roaring lion. While flirting with the lion tamer, she leans over to the side of the cage, and blows smoke in a lion's face, eliciting in response a growl from the lion (Figures 33 & 34). The graphic match functions on two levels: figuratively, the stereotype of a passive plant has been contradicted and the problematic aspect of the mandrake root is revealed: it is an aggressive plant that behaves like an animal. Second, this short visual match extends the threat of Alraune's caged aggression to an entire class. The contained wild animals seen performing beside the circus performers suggests a reading of social outcasts and criminals as a suppressed threat. The graphic match in *Alraune* recalls a similar graphic match in the magazine *Der Querschnitt* from 1926. On opposing pages, a yawning tiger faces a close-up of the inside of an orchid (Figures 35 &

230 I am referring here to Weinstein's argument that the film suggests that Alraune is also racially impure. She cites a scene cut from the final edit that depicted a black man climbing out of the brothel where Alraune's mother was found (Weinstein, 209, note 9).

FIGURE 33 *Lion Roaring from from* Alraune. *Scene still. Dir. Henrik Galeen. Ama-Film, 1928. Illustration reproduction courtesy of the* Friedrich-Wilhelm-Murnau-Stiftung.

36). In bringing the two images together on opposing pages, familiar objects become strange, allowing the orchid to be read as a gaping mouth and the tiger as an open invitation. Similarly, the juxtaposition between Alraune and the lion defamiliarizes Alraune, suggesting that she might be more like the caged lion—contained by the limits of her origins—enforced by her "father" Brinken.

The cages are significant in this scene and throughout the episode with the circus for their representation of ownership and containment as a relationship between classes. At the beginning of the film, the myth warns against attempting to own the mandrake root: "But it could also bring suffering and torment to anyone who tried to own it." A few typical circus acts are followed by Alraune appearing and disappearing from a birdcage. The scene links Alraune's place in the cage to her relationship to Brinken, who watches her from the audience, believing he has caught his wayward bird. On the surface, it appears that Brinken is asserting his control over Alraune. It is a classic example of Laura Mulvey's theory of the male gaze and the objectification of the female body on film. However, after Brinken leaves his seat, Alraune's first appearance in a

FIGURE 34 *Alraune Grimacing from* Alraune. *Scene still. Dir. Henrik Galeen. Ama-Film, 1928. Illustration reproduction courtesy of the* Friedrich-Wilhelm-Murnau-Stiftung.

FIGURE 35
Sam, der Königstiger im Londoner Zoo. From Der Querschnitt. *6 March 1926: 40.* Illustrierte Magazine. *Web. 12 Aug 2013.*

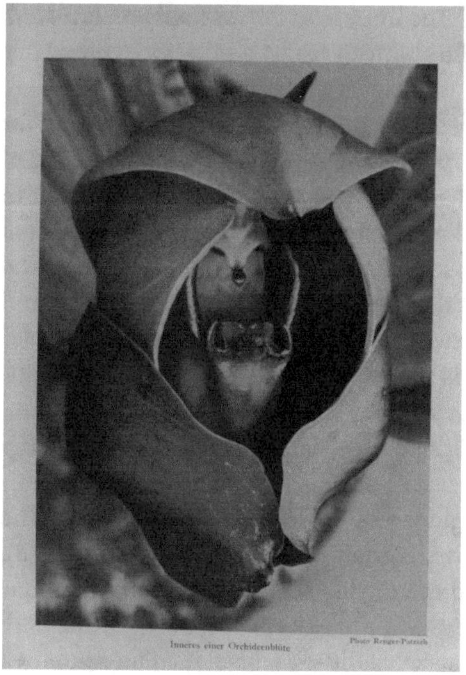

FIGURE 36
Inneres einer Orchidblüte from Albert Renger-Patzsch. From Der Querschnitt. *6 March 1926: 41.* Illustrierte Magazine. *Web. 12 Aug 2013.*

cage is qualified by a second. Entering the lions' cage, she is neither afraid nor tame, but hypnotizes the lions, stopping them in their tracks. People swirl chaotically around the outside of the cage, assuming that she needs rescuing until the lion tamer pushes her out of the cage. He resumes the role that Alraune had overtaken. Here, Alraune plays both the sexual object and the predator, foreshadowing the development of her relationship with Brinken.

The sequence with the circus also reflects thematically on film aesthetics regarding visibility and invisibility, surface appearance and hidden content. The circus performances appear seamless from the perspective of the audience, since the secrets behind the tricks are hidden from their view. The performances recall the early cinema of attractions. As in the films of Georges Méliès, the tricks consist of making objects appear and disappear: a man conjures flowers from an empty vase, and Alraune appears in a cage. But for each performance, the illusion is broken for the film's audience by a shot from behind the stage that reveals the magician's secrets. An equivalent to this balance between illusion and revelation occurs in many of the short plant documentary films. They often mention the rate at which the film had been sped-up or the actual rate of plant growth.[231] The irony here throws doubt on Brinken's ability to read Alraune, as well as the ability of realist aesthetics to reveal her true nature and place in society.

The moments when the realist aesthetic are interrupted are important for determining when Brinken loses his power over Alraune and when their social positions shift relative to one another. One key scene follows her reaction to the truth of her birth using expressionistic stylings to visualize her existential dilemma. In contrast to the bright, low-contrast images from the rest of the film, Alraune's emotional turmoil is depicted using high-contrast lighting and she is wearing a black dress with extended white lapels, long pointed sleeves and hem in the front. After she reads Brinken's journal, a record of his experiment with her, she collapses on the adjacent couch overwrought with emotion and then proceeds to where Brinken is sleeping. The scene repeats the final scene in *Nosferatu*, when he goes to drink Ellen's blood, complete with the looming, oversized shadows (Figures 37 & 38), and the iconic image of the shadow of his hands reaching out over the intended victim (Figures 39 & 40). There is just one exception. Instead of carrying through with her murderous intentions, Alraune restrains herself drawing back instead of acting on her dark urges. With her vampiric emotions successfully restrained, the film resumes a realist aesthetic. Through this episode and the circus, the film undermines the truth-value of New Objectivity by revealing an alternative and unstable reality potentially better suited to discovering Alraune's inner self.

Apart from the visual clues, there are at least two direct references to Alraune's inner state that qualify her humanity and indirectly criticize Brinken's. The first comes in the form of a warning from Brinken's nephew that echoes the common association of warmth with life and emotions: "One who lacks the warmth of life. And bears the chill of death within her." His chilling description identifies her subjective experience as plant-like and therefore inhuman through her coolness and proximity to death. His interpretation is the inversion of Dr. Cinderella's in Meyrink's "Die Pflanzen des Doktor Cinderella," who interprets the warmth of his carnivorous plants to mean that they are animal-like. This description of Alraune could also refer to Brinken at the start of the film, whose character is described in a review from the magazine, *Lichtbild-Bühne*: "One believes of him the scientific cool and the ambivalence towards

231 *Geheimnisses der Pflanzenleben* (*Secrets of Plant Life*) mentions the length of the shot in relation to screen time: "What here in the course of thirty seconds is seen, is in reality a struggle that lasts a full week ("Was hier im Ablauf von dreißig Sekunden sehen, ist in Wirklichkeit ein Kampf, der eine volle Woche dauert").

FIGURE 37 *Nosferatu Motif in Alraune from* Alraune. *Scene still. Dir. Henrik Galeen. Ama-Film, 1928. Illustration reproduction courtesy of the* Friedrich-Wilhelm-Murnau-Stiftung.

humans" ("Man glaubt ihm die wissenschaftliche Kälte und die menschliche Gleichgültigkeit"). The inhuman soul within a mandrake root reflects the monstrous aspect of her creator and of science. It denies her the human love that binds her socially and defines the loveless relationship between her and her father. Her mandrake roots place her outside of society, from the "scum of the Earth" as Brinken specifically requests. His willingness to manipulate humans for the purpose of science reflects his own detachment from the society that appears to support him. When he attempts to murder Alraune at the end of the film, his murderous intentions cause him to resemble more the "scum" of society than the representative of society he once was. Alraune on the other hand strives to be removed from her past by the end of the film. She asks the viscount: "Take me away from here. Give me another soul, and a heart so that I might become a human being, and love like one." Her statement qualifies being human as attaining an inner emotional and spiritual state that builds social relations rather than destroys and is unrelated to the name and social

FIGURE 38 *Nosferatu's Shadow from* Nosferatu. *Scene still. Dir. F. W. Murnau. Minneapolis, Minn.: Mill Creek Entertainment, LLC, 2006. Illustration reproduction courtesy of the* Friedrich-Wilhelm-Murnau-Stiftung.

position. Her marriage contradicts to some extent her definition of human as loving others through its social advantages. Although the film does not definitively show that Alraune will remain with the viscount, it does represent her social climb from an outcast to a titled social member, and the corresponding decline of Brinken from a respected member of society to "Swindler" and finally attempted murderer.

Alraune's reference to the heart brings to mind, *Metropolis* (1927), a film concerned with social instability and class structure. The motto of the film, "in between head and hand is the heart" ("Mittler zwischen Hirn und Hand muß das Herz sein!"), presents brotherly love as a great equalizer that does away with the boundaries that divide class. As in *Metropolis*, a woman, Maria and her robot double, also played by Brigitte Helm, are at the centre of social upheaval that ride on questions of humanness. In *Alraune*, the famous robot from *Metropolis* has been replaced by a plant and artificial breeding, just as the critical look at technology has been replaced by one directed toward science.

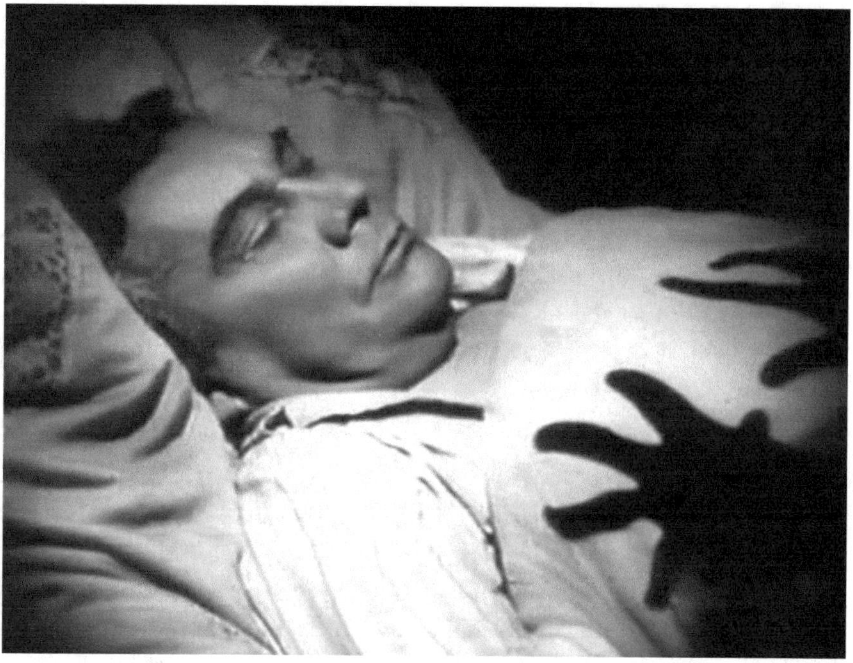

FIGURE 39 *The Shadow of Alraune's Hands on Brinken from* Alraune. *Scene still. Dir. Henrik Galeen. Ama-Film, 1928. Illustration reproduction courtesy of the* Friedrich-Wilhelm-Murnau-Stiftung.

Gender in these two films and others plays a large role in anxieties about social instability,[232] unsurprising considering the changing roles of women during the Weimar republic and their new visibility.

In addition to changing gender roles, the increasing mobility of the population also fuels social instability in *Alraune*. The ability to leave everything behind, including the institutions that once determined your social status, opens the way for a fresh start. Each time Alraune moves to a new place her identity changes. She goes from a schoolgirl at the convent, to an attraction in the circus, to an attractive young woman, to a vamp, and finally to a woman whose choice suggests she has entered into a respectable social position. The first transition is perhaps the most meaningful for Alraune and for the film. Taking place on train, the transition becomes a metaphor for the circulation of people, from all different classes and freely mixing with one another.

232 The examples are numerous and include: *Die Büchse der Pandora* and *Der blaue Engel*.

FIGURE 40 *The Shadow of Nosferatu's Hands on Ellen from* Nosferatu. *Scene still. Dir. F. W. Murnau. Minneapolis, Minn.: Mill Creek Entertainment, LLC, 2006. Illustration reproduction courtesy of the* Friedrich-Wilhelm-Murnau-Stiftung.

In the train, Alraune not only has a sexual awakening symbolized by a shot of the train sliding in between two hills and the mouse running up her skirt, but she also encounters a group of travelling circus performers, whose nomadic lifestyle and place outside of society represents an alternative to hierarchical social relations based on traditional gender roles and titles.[233]

In the film's context of increasing physical and social mobility, the myth of the mandrake root takes on a new meaning. Central and constant in many forms of the myth is the danger involved in taking the root from its place in the ground.[234] When the mandrake root is ripped from the earth, it lets loose a terrible and deadly scream according to most versions of the myth. The

233 The alternative lifestyle of the circus at times was construed as a threat. In an article in the magazine *Das Kriminal-Magzin* from 1931, the author speaks of the reputation of circus performers as unjustly connected to criminality (1667).

234 Gassen and Minol provide a good summary of the changing myth of the mandrake and its role in different historical contexts.

separation of the root from the earth resembles a traumatic and violent experience through the mandrake's scream. Whether the scream is of anger or pain, the consequences are nonetheless deadly. The subsequent potential of the root to bring luck or misfortune introduces an unstable element that is reflected in *Alraune*. Also, the sense of rootlessness and the trauma associated with removing the root from the earth are central themes in *Alraune*. Rootlessness and social mobility are visualized in the last shot of the mandrake root against an abstract background, which once again is shown rotating and then dissolving into Alraune. The plant becomes a woman as easily as Alraune seems to discard her past.

The film, *Alraune*, seems to illustrate a shift in the perception of social stability in Weimar culture that occurred after the economic and political stabilization that began around 1924. In contrast to *Nosferatu*, where the instability has become naturalized, in *Alraune*, the increasing social stability encourages the perception of social mobility and changing gender roles as unnatural. The narrative of a plant becoming human functions as an extended metaphor for underlying anxieties surrounding the inability to discern social class from appearance in addition to fears of the increasing public presence and power of women. Although Alraune appears to be the typical member of the bourgeoisie, her appearance masks her origins. Alraune embodies the many different groups that compose the lower classes and outcasts from criminals, prostitutes or public women, racial minorities, and the circus performers. In the stabilization period of the Weimar republic, Alraune represents anxieties that all is not what it seems. By labelling Brinken's experiments as unnatural, Alraune's apparent ascent and assimilation into the "right circles" goes by association against social rules. Brinken's descent into criminality is by the same token punishment for experimenting with the rules for determining social class.

In contrast, *Nosferatu*, represents social upheaval as a natural event resulting from a cataclysmic force that science and other social institutions are powerless to prevent. In both films, however, questions of social instability are grafted onto representations of plants that defy their traditionally assigned position in the "chain of being." In *Nosferatu*, the animalistic Venus flytrap, and the plant-like polyp, serve to naturalize fluidity, instability and transgression as a general state of nature and society: a "unifying principle" stretching from the microscopic world of the polyp to the upheaval wrought in Wisborg by the vampire. They reveal the film's terrifying proposition that the vampire-as-other is only a façade: that vampirism is ubiquitous and is in fact the norm rather than an outside. In *Alraune*, on the other hand, the promiscuity of a mandrake root becoming human—along with Alraune's continued proximity to the animal world—is represented as an act of deviance, for which the scientist Brinken must be punished at the end of the film. What these films share is the representation of fluidity between plant, animal and human worlds as something terrifying. Whereas a film like *Blumenwunder* could tap into the desire for a child-like experience of the world, in which plants, animals and people could exist in a tender visual proximity, *Nosferatu* and *Alraune* represent the loss of clear boundaries as a kind of nightmare vision, where the stability of clearly distinguished levels of being threatens to descend into chaos. If the former represented the utopian promise of film, the latter registered the anxieties of a society wrought by fears of social change.

Conclusion

> The lowest animal, in this sense, is wholly comparable with the highest plants. The difference, which at first glance causes the animals to appear as living and the plants as lifeless, is due only to the *tempo* of events. All reaction movements are quicker in animals. The movements of flowers have been photographed and transferred to a cinematograph, and then reproduced in the *tempo* of animal movements. They gave the fantastic picture of some fabulous being in tremendous agitation.
>
> FRANCÉ, *Germ of Mind in Plants*, 144[235]

∵

In 1905, Raoul Francé remarked on the capacity of film to depict plant movement, noting the animal-like movements of plants revealed by these short time-lapse films. He recognized that films could help illustrate his concept of a dynamic natural world that contrasted with traditional botanists' practice of accumulating plant specimens, drying and labelling them. His remark points to two broad transformations at the turn of the nineteenth century that placed into question the classification, definitions, and hierarchies of life formed out of the Aristotelian and Judeo-Christian traditions. The first was intellectual—in reaction to what was seen as the excessive materialism and positivism of the mid-nineteenth century, many artists, scientists and writers turned to aspects of Romanticism and Vitalism. This intellectual climate provided the right conditions for a revival of Goethe's explorations of a dynamic nature and Gustav Fechner's speculations on a world soul; both had been ignored by their respective contemporaries in the scientific community, but found new resonances

235 "Das niedere Tier ist in diesem Sinne der hochentwickelten Pflanze völlig gleich. Der Unterschied, der unserem ersten Blick das Tier als lebend und die Pflanze als unbelebt erscheinen läßt, ist ur das Tempo der Vorgänge. Alle Bewegungsreaktionen spielen sich beim Tiere eben rascher ab. Man hat die Nutationen eines Pflanzensprosses kinematogrphaisch aufgenommen und dann in dem Tempo einer tierischen Bewegung reproduziert. Sie gaben das phantastische Bild eines in ungeheurer Aufregung um sich schlagenden Fabelwesens" (Francé, *Das Sinnesleben der Pflanzen*, 86).

 | DOI 10.1163/9789004327177_008

and new interpretations in the post-Darwinian world. This intellectual climate overlapped with the emergence of new forms of time-based media that allowed plant movement to be visualized for the first time. Situated in this context, the "dynamic plant" came to form a privileged motif, which illustrates a shift from classifying nature as a set of discrete objects within a materialist and mechanical perspective to one of processes, intuition and unity. New technologies such as film helped to visualize this "other" nature otherwise invisible to the naked eye. As the preceding chapters have shown, the dynamic plant stands at the heart of these two broad transformations, illustrating the historical intertwining of technology with ideas.

In order to investigate the cultural significance of the dynamic plant in German modernity, six representative works from literature and film were chosen as case-studies for close readings divided in four chapters. The literary examples were Kurd Lasswitz's *Sternentau* (1909), Paul Scheerbart's "Flora Mohr: Eine Glasblumen-Novelle" (1909) and Gustav Meyrink's "Die Pflanzen des Doktor Cinderella" (1905). The primary films were *Das Blumenwunder* (1926), *Nosferatu* (1921) and *Alraune* (1928). I found that the examples could be placed along a spectrum from positive to negative reactions to the concept of a dynamic plant. At one end, plant movement was seen—in the tradition of romantic philosophers and scientists such as Fechner—as the trace of a creative force running throughout the universe and as an opportunity to understand and experience a radically different way of being. At the other end, the destabilization of traditional hierarchies that plant movement implied was seen as subversive and tapped into cultural anxieties about social stability and the body.

The first chapter concentrates on Kurd Lasswitz's model of plant-human communication as derived from Neo-Romantic, Vitalist and scientific discourses at the *fin de siècle*. In his early science fiction novel, *Sternentau*, Lasswitz bridges the gap between humans and plants through the use of the "Universal Translator," an imaginary medium based on the humanoid bodies of an alien species. The alien Idonen solve inter-species communication through forming a network of communication that addresses the barriers, which block acceptance of a plant consciousness. Even though the Idonen eventually depart the Earth, partially because of the poor treatment they experienced at the hands of the scientist, they leave behind a suggestive legacy. Lasswitz novel concludes that by combining the slow methodical practice of science with the imagination of a poet, the interior state of other non-human, living beings can be eventually experienced. In essence, his novel is his attempt to demonstrate this practice, when he integrates the contemporaneous discoveries of the connection between nerves and the mind, the Romantic trope of the blue flower as the representation of consciousness and the emphasis on life as a process of self-actualization.

In contrast to the first chapter, the second chapter reflects on Paul Scheerbart's short story as a celebration of the technical and intellect achievements of humans to enhance nature. Artificial and mechanically animated, the fantastical glass plants at the centre of his short story, "Flora Mohr," are paradoxically an expression of "life" and "soul." Scheerbart's spectacular flower gardens harken back to 19th-century visual spectacles such as the panorama and technology exhibitions at World Fairs, but rather than representing outward appearances, Scheerbart's mechanical plants sought (like Loïe Fuller's light and electricity dances) to imitate the *dynamism* of nature. Drawing on the ideas of Goethe and the Romantics, Scheerbart thus transforms industrial technology—an icon of materialism—into the vitalistic expression of a creative life force.

The perception of the plants as animated living beings and the enchantment of technology link Scheerbart's glass plants and Lasswitz's hybrid Idonen with the focus of chapter three, the film *Das Blumenwunder* (1926). Assembled from a collection of time-lapse films of moving plants and clips of flower and plant dances, the film interprets the accelerated plant movement as evidence of life, transforming in the process the technology of film into a means of accessing the creative force of nature. Through the juxtaposition of plant movement and dancers, in addition to the foregrounding of film techniques such as time-lapse and montage, film becomes an interactive medium intended to teach empathy with plants through mimesis. The film is an exercise in building associations and diffusing distinctions between plants and humans. As with Scheerbart's short story, *Das Blumenwunder* also presents the complexity and movement of visual images as better suited to express life than language. Unlike the categorical thinking encouraged by verbal language, visual media, motion pictures, according to *Das Blumenwunder*, can facilitate a mimetic relation based on the recognition of similarities rather than distinctions.

In contrast to *Sternentau*, "Flora Mohr" and *Das Blumenwunder*, Gustav Meyrink's "Die Pflanzen des Doktor Cinderella" (1905) portrays animated plants as a destructive force, using the concept of a carnivorous plant to critique occultist practices, and materialism within the medical sciences for their Cartesian dualism of the mind and body. Occultist practices are criticized for inducing a state through ecstatic imitation where the soul leaves the body, reducing it to mere matter—to a moving, sensing and acting plant. Here, the occultist practices overlap with the sciences, which are critiqued for viewing the world as mere matter (or body parts) to be exchanged freely as one would exchange parts of the machine without changing the nature of the being. Dr. Cinderella's animated carnivorous plants are a result of a chain of events that began with the occult and ended with gruesome medical experiments in a dark laboratory. Using human and animal body parts in place of leaves, berries and stems, Dr. Cinderella produces de-evolved humans and animals—reduced to the base function of plant growth. The resulting demonic plants subvert an underlying hierarchical order that places humans at the top, followed by animals and then plants, revealing, in the process, the fragile state of the unity of self and of the human being's position in relation to the natural world.

The demonic, carnivorous plant reappears in the two Weimar era films, *Nosferatu* and *Alraune*, as a reflection of the cultural anxieties surrounding the "foreigner" and the "new woman" after WWI. As with Meyrink's short story, both films blend the occult with science and technology to produce a monstrous animated plant. In *Nosferatu*, the carnivorous Venus flytrap becomes a metaphor for the way the intrusion of a foreigner into Weimar culture destabilizes the traditional hierarchies embodied (at least for Hutter) by the flowers at the start of the film. In *Alraune*, the blending of mandrake root and human produces a protagonist who appears to move freely between social classes. Measured against the traditional metaphor of the woman as flower, Alraune defies gender conventions in a similar manner to how the animated plant subverts the common expectation of plant behaviour.

Originally, I intended to focus this book on time-lapse films. Many of these films from the 1910s and 1920s were unavailable for viewing for various reasons—some have been lost and others exist only as copies of the extremely flammable nitrate base. But the few films I did view—including gems such as *Die Seele der Pflanze* (1922), *Pflanzen leben* (1920s) and *Das Blumenwunder* (1926)—pointed to a larger cultural preoccupation with the moving plant. These films made the slow movement of plants visible for anyone to see—a dynamism that has been often overlooked in face of the apparent inertia of plants in any single instance. I believe that the time-based films contributed to the reappearance of a reoccurring discussion surrounding the basis of plant life. One of the fundamental points of the discussion involves the nature of the plant as a living being. These films made visible the animal-like movements of plants, proving unequivocally for the viewers that plants are living and possibly sentient beings. In the chain of being that persists during this period plants are placed just above crystals and other mineral matter, suggesting that in the human perception of the natural world, plants are closer to inanimate objects than animals. The lack of attention to plants' movement allowed for them to be read as closer to inanimate objects. The time-lapse films of plants moving helped to change this perception of plants, making visible similarities between animal and plant movement—similarities further emphasized in the films I examined through the use of associative montage.

By connecting the motif of the dynamic plant to the intellectual climate as well as new media, this book contributes to a growing field of research that addresses the relative neglect and denial of discourses around nature at the *fin de siècle*, as exemplified in the 2011 anthology, *Biocentrism and Modernism*. Edited by Oliver Botar and Isabella Wünsche, the anthology identifies a series of discourses, which they call "biocentrism." While diverse and distinct, all of these discourses share a set of beliefs characterized by aspects of Neo-Romanticism and Neo-Vitalism. Although the goal of my research has not been to identify the films and short stories as biocentric (but rather to uncover the reactions to the transformations in ideas and technology), they certainly share many affinities with Botar and Wünsche's brief definition of biocentrism as "Nature Romanticism updated by the Biologism of the mid- to late nineteenth century" (2). Because of the affinities, many of the same names that Botar and Wünsche associate with biocentrism, notably Raoul Heinrich Francé, also appear in association with the motif of a dynamic plant. Since biocentrism shapes their anthology as a historical concept rather than a coherent school or a movement, the concept also helps to explain the diversity of films and short stories chosen for this book that range from canonical (*Nosferatu*) to relatively obscure (*Sun-Dew, The Miracle of Flowers*) and in styles from fantasy ("Flora Mohr") and horror ("Doctor Cinderella's Plants" to expressionism and new objectivity (*Alraune*).

This interest in the relationship between media and life at the beginning of the 20th century is echoed by a new kind of vitalistic thinking in film and media studies at the turn of the 21st century—one that sees media as able to convey the dynamic life processes of the natural world but in a sense also a part of that "life." Thus Sarah Kember and Joanna Zylinska refer to the "*vitality of media*," insisting on "our entanglement with media not just on a socio-cultural, but also on a biological level" (xvii–xviii). This view of media finds something of a precursor in Scheerbart's short story, "Flora Mohr," where the fantastical glass plants appear to foreshadow this doubled perception of media as representing the process of a living being and somehow also alive. Scheerbart's perception of media as life forms resonates with examples from Weimar film listed in Jussi Parikka's *Insect Media* (2010). Parikka argues that insects perform media and media can be also perceived as a life form, citing the German language film, *Die Biene Maja und ihre Abenteuer* (*Maya the Bee and her Adventures*) (1924) as an early example of how insects were a lesson in "rational social organization" (42).[236] Parikka's specific focus on insects resonates with a broader recent tendency toward viewing and analyzing media as an ecology, which, in part, seeks to address the relations between material and immaterial aspects of media. One of the more seminal studies comes from Matthew Fuller, whose comprehensive book, *Media Ecologies* (2005), covers a range of topics from pirate radio to film with a reflective look at the term "media ecology."

The view of media as life is often accompanied by the need to define consciousness, will or even soul. Part of Fuller's *Media Ecologies* is engaged with examining the "will to power" in media systems. By making visible plant movement, the early time-lapse films also helped to revive another reccurring debate surrounding the nature of plants: whether they possess a mind and soul. Since Aristotle theorized that there are degrees of soul—the most basic being plants, then animals and ending with humans, the idea that plants may have a soul has repeatedly resurfaced. Drawing on the natural philosophy movement of Romanticism, Gustav Fechner expanded the definition of soul to include the plants and the cosmos in the mid 19th century. His ideas, at first ignored or derided by the scientific community, enjoyed a revival at the beginning of the 20th century with writers such as Kurd Lasswitz and Paul Scheerbart, botanists like Raoul Francé, scientists such as Ernst Haeckel and educators like Rudolf Steiner. While resonating with vitalism, this insistence that plants may have a kind of consciousness and even a soul was also bound up with currents of environmental thought that emerged around 1900, which promoted deep sense of respect for the natural environment. This environmental message also clearly underlies the recent interest in the perception and being of plants in philosophy, represented respectively by Matthew Hall's book, *Plants as Persons* (2011) and Michael Marder's *Plant Thinking* (2013).

In tracing the effect of media on our perception of the plants in the early 20th century, this book contributes to a growing body of scholarship within German studies that is concerned with questions of environment and culture. Classical ecocriticism focuses on the environmental imagination (Buell, *The Environmental Imagination*), literature and the environment (Glotfelty), the green German tradition (Hermand) and the culture of German environmentalism (Goodbody, *The Culture of German Environmentalism*). More recent books have expanded the boundaries from nation and language to the global

236 Other examples he cites include the Weimar films, *Metropolis* (1927) and *Nosferatu* (1921). The first he compares to an ant colony with workers on the bottom and a queen on top (42). The second, he notes, translates between images and between the human body into other species (95).

imagination (Heise), looked to the future of ecocriticsm (Buell, *Future of Environmental Criticism*), and sought to deepen its theoretical base (Goodbody and Rigby, *Ecocritical Theory*). Ecocritical approaches to literature highlight its role in shifting the perception of the natural world from the Western standpoints that encourage an increasing alienation from nature to an understanding of the self as embedded in the surrounding environment (Wilke 170). To some extent, all six primary works discussed in this book lend themselves to an ecocritical approach. But the two most explicit attempts to bring about a change in the perception of nature can be found in Lasswitz's *Sternentau*, and *Das Blumenwunder*. Lasswitz's novel explicitly connects the treatment of plants and the alien Idonen in the interest of science with the perception of plants as objects. *Das Blumenwunder* represents aesthetic experience as a means to teach bodily communication with plants and respect of them as living beings. Scheerbart's utopian short story, "Flora Mohr," is an attempt to change the relationship of visual art to the natural world, using technology to replicate the flows and processes of a dynamic world rather than merely copying the appearance of plants as static objects. The subtexts of the remaining three works—Meyrink's short story, "Die Pflanzen des Doktor Cinderella," *Nosferatu* and *Alraune*—acknowledge the loss of traditional hierarchies responding negatively to the intellectual and social instability that loss implies.

Although Lasswitz's novel *Sternentau*, Scheerbart's short story, "Flora Mohr," and Meyrink's "Die Pflanzen des Doktor Cinderella" take an explicit stand against materialism, all of the works foreshadow to some extent what has become known as "new Materialism" in recent theory, as articulated in the Diana Coole and Samantha Frost's collection, *The New Materialisms* (2010). This perspective is characterized, in Coole and Frost's formulation, by a belief that "materiality is always something more than 'mere' matter: an excess, force, vitality, relationality, or difference that renders matter active, self-creative, productive, unpredictable" (9). Derived from the vitalist thinkers Gilles Deleuze and Henri Bergson, the approach recognizes the "realness" of matter that is also not a sum of its parts as in 19th century materialism. This view of matter resonates with Meyrink's dismissal of the machine model of plants as well as his rejection of occult practices that eschewed the body. His example shows that the rejection of materialism as it was understood at the beginning of the 20th century was not necessarily founded on a rejection of the material world, but rather on a search for a model of being based on the complexity of life unfolding through time. The examples of literature and film discussed in this book all grapple with the perception of a natural world that consists not of inert objects and stabilized social structures, but of dynamic plants that uncover the social and natural flux that characterizes any relationship.

There would be many directions for further research based on this book. One possibility would be to return to the early time-lapse films of plants and compare them to more contemporary films in order to find out how nature films have changed their representation of plants. Great strides into answering this question have already been made in *Perspektiven einer anderen Natur* by Andreas Becker, whose survey of time-lapse photography and slow motion films on nature is both comprehensive and insightful. He traces the origins of time-lapse films with plants to German botanist, Wilhelm Pfeffer's, experiments with photographing tulips in 1898, and follows the development through early German *Kulturfilm*, Weimar cinema, Disney's nature films and David Attenborough's well-known nature documentaries. While narrower in scope, Oliver Gaycken's article "The Secret Life of Plants: Visualizing Vegetative Movement, 1880–1903" (2012) outlines the early developments of time-lapse photography of plants in Germany, and demonstrates the impact they had on the perception of plants in popular culture. As Gaycken's and Becker's research has demonstrated, the scientifically coded image of plants growing has made its way into the broader cultural understanding of plants. Through the discussion of the films, *Das Blumenwunder* and *Nosferatu*, this book has begun to examine the way scientific images of plants have been integrated into mainstream feature films. The interpretations of moving plants in the two films differ greatly from one another and provide a glimpse into the cultural meanings of the dynamic plant as well as the cultural significance of certain plants. One possible avenue for further research would be to examine the way images of plants are integrated into contemporary mainstream features.

Another direction for further research would be to examine representations of dynamic plants within a broader selection of modern literature, extending beyond the German-speaking world. Especially within the science fiction genre, many examples can be found of the dynamic plant motif. In his short story, "The Flowering of the Strange Orchid" (1894), the well-known English writer, H. G. Wells featured a blood-sucking orchid with tentacles, which grabs onto its victim. His story resonates with Meyrink's use of orchid imagery to depict of the vampiric tendencies of a beautiful woman in his short story, "Bologneser Tränen." Another science fiction writer, the American, Howard Garis, wrote a short story, "Professor Jonkin's Cannibal Plant" (1905) featuring a carnivorous plant that attempts to eat its breeder, Professor Jonkins. These examples of vampiric plants indicate a trend in early science fiction of the demonic plant that is worth further exploration.

This book comes at a particularly exciting time for German and media studies. Not only are new fields of study, such as ecocriticism, being developed to address pressing real-life concerns about the sustainability of life on this planet, the ethics of our relationship to plants, and our estrangement from the natural world, but also the role of media as the "other nature" is being examined for its impact on the environment and for how it shapes the way we perceive and interact with the natural world. By writing on plants in films and short stories at the turn of the century, I am tracing the way concepts and ideas impact our perception of plants, but also examining the way an intellectual climate is reflected in or has an influence on media and artistic forms, such as film, dance and glass sculpture to encourage certain interpretations over others. As represented in the motif of the dynamic plant, the variety of reactions to the intellectual and technological transformations at the turn of the century has provided a glimpse into the complexity of the relations between media and ideas—encompassed by a blend of resistance and acceptance.

Bibliography

Abram, David. *The Spell of the Sensuous: Perception and Language in a More-than-Human World*. New York: Vintage Books, 1996. Print.

Adelt, Leonhard. "Lesabéndio. Ein Asteroiden-Roman." Rev. of *Lesabéndio. Ein Asteroiden-Roman* by Paul Scheerbart. *Das litterarische Echo*. 2 Jan. 1914. 650–52. *Über Paul Scheerbart*: Ed. Paul Kaltefleiter. Oldenburg 1998. 221–23. Print.

Adorno, Theodor and Max Horkheimer. *Dialektik der Aufklärung: Philosophische Fragmente*. Frankfurt a.M.: Fischer, 2008. Print.

Alcock, John. *An Enthusiasm for Orchids: Sex and Deception in Plant Evolution*. Oxford: Oxford University Press, 2006. Print.

Alraune. Dir. Henrik Galeen. Perf. Brigitte Helm and Paul Wegener. Ama-Film, 1928. Film.

"Alraune." Rev. of *Alraune*. *Lichtbild-Bühne*. Feb. 1928. *Filmportal*. Web. 12 Aug. 2013.

Andriopoulos, Stefan. "Occult Conspiracies: Spirits and Secret Societies in Schiller's Ghost Seer." *New German Critique; Ngc*. (2008): 65. *ArticleFirst*. Web. 10 Dec. 2014.

Arnheim, Rudolf. "Das Blumenwunder (1926)" from *Film als Kunst. Presse über den Film "Das Blumenwunder". Archiv der Deutschen Kinemathek*. Print.

The Arabian Nights' Entertainments. Chicago: Rand McNally and Co., 1914. *Project Gutenberg*. Web. 24 Apr. 2012.

Asendorf, Christoph. *Batteries of Life: The History of Things and Their Perception in Modernity*. Trans. Don Reneau. *Weimar and Now: German Cultural Criticism*. Ed. Martin Jay and Anton Kaes. Berkeley: U of California P, 1993. *Google Books*. Web. 22 Jul. 2012.

Bakke, Monika. "Art for Plant's Sake? Questioning Human Imperialism in the Age of Biotech." *Parallax*. 18.4 (2012): 9–25.

Balázs, Béla. "Tanzdichtungen" *Musikblätter der Anbruch*. March 1926: 109–112. *Internet Archive*. Web. 2 Dec. 2013.

Balázs, Béla. *Der Sichtbare Mensch oder die Kultur des Films*. (1924). Frankfurt am Main: Suhrkamp, 2001. Print.

Bär, Hubert. *Natur Und Gesellschaft Bei Scheerbart: Genese Und Implikationen Einer Kulturutopie*. Heidelberg: J. Groos, 1977. Print.

Becker, Andreas. *Perspektiven Einer Anderen Natur: Zur Geschichte Und Theorie Der Filmischen Zeitraffung Und Zeitdehnung*. Bielefeld: Transcript, 2004. Print.

Becker, Erich. *Deutsche Philosophen: Lebensgang u. Lehrgebäude von Kant, Schelling, Fechner, Lotze, Lange, Erdmann, Mach, Stumpf, Bäumker, Eucken, Siegfried Becher*. München: Duncker & Humblot 1929. Print.

Beerling, D. J. *The Emerald Planet: How Plants Changed Earth's History*. Oxford: Oxford University Press, 2007. *Ebrary*. Web. 30 May 2013.

Behne, Adolf. *Die Wiederkehr der Kunst*. Leipzig: K. Wolff, 1919. *Internet Archive*. Web. 28 Dec. 2011.

Benjamin, Walter. *Das Kunstwerk im Zeitalter seiner technischen Reproduzierbarkeit*. 3rd Version. *Gesammelte Schriften*. Ed. Rolf Tiedemann, and Hermann Schweppenhäuser. Vol. 1. Frankfurt am Main: Suhrkamp, 1978. 371–508. Print.

Benjamin, Walter. "Doctrine of the Similar." 1930. *New German Critique*. (Spring 1979). 65–69. *Jstor*. Web. 17 August 2011.

Benjamin, Walter. "Paris, die Stadt im Spiegel." *Gesammelte Schriften*. Ed. Rolf Tiedemann, and Hermann Schweppenhäuser. Vol. 4. Frankfurt am Main: Suhrkamp, 1978. 356–58. Print.

Benjamin, Walter. "Kleine Geschichte der Photographie." (1931) *Gesammelte Schriften*. Ed. Rolf Tiedemann, and Hermann Schweppenhäuser. Vol. 2. Frankfurt am Main: Suhrkamp, 1978. 368–86. Print.

Benjamin, Walter. "Neues von Blumen" Rev. of *Urformen der Kunst* by Karl Blossfeldt. *Die literarische Welt*. 23 Nov. 1928. *Gesammelte Schriften*. Ed. Rolf Tiedemann, and Hermann Schweppenhäuser. Vol. 3. Frankfurt am Main: Suhrkamp, 1978. 151–53. Print.

Benjamin, Walter. "Zu Scheerbart "Münchhausen und Clarissa." *Gesammelte Schriften*. Ed. Rolf Tiedemann, and Hermann Schweppenhäuser. Vol. 6. Frankfurt am Main: Suhrkamp, 1978. 148–49.

Bergson, Henri. *Creative Evolution*. 1911. Trans. Arthur Miller. New York: Modern Library, 1944. *Internet Archive*. Web. 9 April 2014.

Bergstrom, Janet. "Sexuality at a Loss: The Films of F. W. Murnau." *Poetics Today*. 6 (1985): 185–203. Print.

"Berliner Morgenpost." *Presse über den Film "Das Blumenwunder"*. Archiv der Deutschen Kinemathek.

Blankenship, Janelle. "'Film-Symphonie vom Leben und Sterben der Blumen': Plant Rhythm and Time-Lapse Vision in Das Blumenwunder". *Intermédialités: histoire et théorie des arts, des lettres et des techniques / Intermediality: History and Theory of the Arts, Literature and Technologies*, n° 16, 2010, 83–103.

Bletter, Rosemarie Haag. "Bruno Taut and Paul Scheerbart's Vision: Utopian Aspects of German Expressionist Architecture." Diss. Colombia U. 1973. *ProQuest Dissertions and Theses*. Web. 22 Aug. 2012.

Bletter, Rosemarie Haag. "The Interpretation of the Glass Dream-Expressionist Architecture and the History of the Crystal Metaphor." *Journal of the Society of Architectural Historians*. 40.1 (1981): 20–43 *Jstor*. Web. 28 Nov. 2011.

Blossfeldt, Karl. *Adiantum pedatum. Frauhaarfarn*. 1928. *Karl Blossfeldt: The Complete Published Work*. Köln: Taschen, 2008. 79. Print.

Blossfeldt, Karl. *Urformen der Kunst*. 1926. *Karl Blossfeldt: The Complete Published Work*. Köln: Taschen, 2008. 124–135. Print.

Boewe, Karl-Heinz. “Paul Scheerbart: Romanthemen und Erzähltechnik” Diss. Rice U. 1969. *ProQuest Dissertations and Theses*. Web. 30 Aug. 2012.

Botar, Oliver A. I., and Isabel Wünsche. *Biocentrism and Modernism*. Burlington, VT: Ashgate, 2011. Print.

Boyd, Amanda. “Gustav Meyrink and the Evolution of the Literary Vampire: from Feared Bloodsucker to Esoteric Phenomenon.” *Neophilologus*. 90.4 (2006): 601–620. Print.

Buell, Lawrence. *The Future of Environmental Criticism: Environmental Crisis and Literary Imagination*. Malden, MA: Blackwell Pub, 2005. Print.

Buell, Lawrence. *The Environmental Imagination: Thoreau, Nature Writing, and the Formation of American Culture*. Cambridge, MA: Belknap Press of Harvard University Press, 1995. Print.

Brunzlow, Herbert. *Über die Anwendung psychologischer Kategorien auf Pflanzen bei Fechner und Francé*. Breslow. Dissertation. 1920.

Calhoon, Kenneth S. “F. W. Murnau, C. D. Friedrich, and the Conceit of the Absent Spectator.” *Mln*. 120.3 (2005): 633–653. Print.

Chamovitz, Daniel. *What a Plant Knows: A Field Guide to the Sense*. New York: Scientific American/Farrar, Straus and Giroux, 2012. Print.

Clason, Christopher. “Automatons and animals. Romantically manipulating the chain of being in E. T. A. Hoffmann’s ‘Der Sandmann’ and ‘Kater Murr.’ ” *Romanticism and Beyond: A Festschrift for John F. Fetzer*. Ed. John F. Fetzer, Clifford A. Bernd, Ingeborg Henderson, and Winder McConnell. New York: P. Lang, 1996. 115–132. Print.

Cohen, Jeffrey Jerome. “Monster Culture (Seven Theses).” *Monster Theory: Reading Culture*. Ed. Jeffrey Jerome Cohen. Minneapolis: Minnesota Press, 1996. *Ebrary*. Web. 15 Apr. 2013.

Cowan, Michael. “Cutting through the Archive: Querschnitt Montage and Images of the World in Weimar Visual Culture,” *New German Critique* 120, vol. 40, no. 3 (2013). 16–21. *Crossref*. Web. 20 Dec. 2013.

Cowan, Michael. *The Cult of the Will: Nervousness and German Modernity*. University Park, Pa: Pennsylvania State University Press, 2008. Print.

Creutz, Max. “Paul Scheerbart” *Westdeutsche Wochenschrift* 1919: 48. *Über Paul Scheerbart: 100 Jahre Scheerbart-Rezeption: in Drei Bänden*. Ed. Michael M. Schardt. Vol. 3. Paderborn: Igel, 1992. 394–96. Print.

“Dancing with Plants” Science Museum of Minnesota: St. Paul. Web. 6 Jan. 2014.

Darwin, Charles. *Insectivorous Plants*. Ed. Francis Darwin. 2nd ed. London: John Murray, 1888. *Internet Archives*. Web. 31 Aug. 2013.

Darwin, Charles. *The Power of Movement in Plants*. New York: Appleton, 1895. *Internet Archive*. Web. 6 Sept. 2013.

Darwin, Charles. *The Movements and Habits of Climbing Plants*. 2nd ed. London: Murray 1875. *Google Books*. Web. 6 Sept. 2013.

Das Blumenwunder. BASF AG (Ludwigshafen); Unterrichtsfilm GmbH, Verlag wissenschaftlicher Filme (Berlin), 1922–26. *Filmarchiv-Bundesarchiv*. Film.

Das Blumenwunder: Ein Film. Film program. 1926. *Archiv der Deutschen Kinemathek. Arte.tv*. Web. 18 Nov. 2013.

Daston, Lorraine, and Peter Galison. "The Image of Objectivity." *Representations*. (1992): 81–128. Print.

Daston, Lorraine. *Objectivity*. New York: Zone Books, 2007. Print.

The Cabinet of Dr. Caligari: A Film in Six Acts. Dir. Robert Wiene. New York, NY: Kino on Video, 2002.

"Der Montag Morgen." *Presse über den Film "Das Blumenwunder". Archiv der Deutschen Kinemathek*. Print.

Die Seele der Pflanzen. Dir. Max Planck. Universum Film, 1922. *Filmarchiv-Bundesarchiv*. Film.

Die Seele der Pflanze. Dir. Max Brink, UFA, 1922. Censorship card. *Filmarchiv-Bundesarchiv*. Microfilm.

Döblin, Alfred. "Die Ermordung einer Butterblume." *Die Ermordung einer Butterblume: Und Andere Erzählungen*. München: G. Müller, 1913. *Internet Archive*. Web. 14 Oct. 2014.

Dörfel, Günter, and Falk Müller. "1857—Julius Plücker, Heinrich Geissler Und Der Beginn Systematischer Gasentladungsforschung in Deutschland." *Ntm International Journal of History and Ethics of Natural Sciences, Technology and Medicine*. 14.1 (2006): 26–45. Print.

Doxey, Denise M. "Anubis." *The Oxford Encyclopedia of Ancient Egypt*. 2001. *Oxford Reference*. Oxford: Oxford University Press, 2002. Web.

Dulac, Germaine. "The Essence of the Cinema: The Visual Idea." Trans. Robert Lamerton. Ed. P. A. Sitney *The Avant-Garde Film: A Reader of Theory and Criticism*. New York: New York University Press, 1978. 36–42. Print.

Dulac, Germaine. *Thèmes et Variations*. 1928. Film.

Durkheim, Émile, and Kenneth Thompson. *Readings from Emile Durkheim*. London: Routledge, 2004. *Milibrary*. Web. 9 July 2013.

Eisner, Lotte H. *The Haunted Screen: Expressionism in the German Cinema and the Influence of Max Reinhardt*. Berkeley: University of California Press, 1973. Print.

Ege, Müzeyyen. *Das Phantastische Im Spannungsfeld Von Literatur Und Naturwissenschaft Im 20. Jahrhundert: Die Pluralität Der Welten Bei Paul Scheerbart, Carlos Castaneda Und Robert Anton Wilson*. Berlin: wvb, Wiss. Verl, 2004. Print.

Erb, Wilhelm. *Über die wachsende Nervosität unserer Zeit*. Heidelberg: Universitäts-Buchdruckerei von J. Hörning, 1893. *Internet Archive*. Web. 11 Jan. 2014.

Fechner, Gustav. *Nanna; Oder, Über Das Seelenleben Der Pflanzen*. Leipzig: L. Voss, 1848. *Google Books*. Web. 17 Sept. 2013.

Fechner, Gustav. *Zend-avesta, Oder, Über Die Dinge Des Himmels Und Des Jenseits*. Leipzig: Lepold Voss, 1851. *Google Books*. Web. 18 Sept. 2013.

Fiedler, Kuno. *Die Motive der Fechner'schen Weltanschauung*. Leipzig, Phil. Diss., 1918.

Fischer, William. *The Empire Strikes Out: Kurd Lasswitz, Hans Dominik, and the Development of German Science Fiction*. Bowling Green, Ohio: Bowling Green State University Popular Press, 1984. Print.

Fleishfressende Pflanzen. Deulig, 1922. Censorship card. *Filmarchiv-Bundesarchiv*. Microfilm.

Fleischfressende Pflanzen. Dir. Dr. Ulrich Schulz. Ufa, 1943. *Filmarchiv-Bundesarchiv*. Film.

"Filmkurier" *Presse über den Film "Das Blumenwunder"*. Archiv der Deutschen Kinemathek. Print.

Foster, Susan. *Choreographing Empathy: Kinesthesia in Performance*. New York: Routledge, 2011. Print.

Francé, Raoul Heinrich. *Das Sinnesleben Der Pflanzen*. Stuttgart: Kosmos gesellschaft der naturfreunde, 1905. Print.

Francé, R. H., and Algie M. Simons. *Germs of Mind in Plants*. Chicago: C. H. Kerr & Co, 1905. *Internet Archive*. Web. 7 Jan. 2013.

Freud, Sigmund. "Das Unheimliche." *Imago: Zeitschrift für Anwendung der Psychoanalyse auf die Geisteswissenschaften*. 5/6 (1919): 297–324. *Internet Archive*. Web. 3 Jan. 2014.

Fuller, Loië. *Fifteen Years of a Dancer's Life: With Some Account of Her Distinguished Friends*. London: H. Jenkins Ltd, 1913. *Internet Archive*. Web. 27 Sept. 2013.

Fuller, Matthew, and Roger F. Malina. *Media Ecologies: Materialist Energies in Art and Technoculture*. Cambridge, Mass.: MIT Press, 2005. *Ebsco*. Web. 5 Jan. 2014.

Gaycken, Oliver. "The Secret Life of Plants: Visualizing Vegetative Movement, 1880–1903." *Early Popular Visual Culture*. 10.1 (2012): 51–69. Print.

Gaycken, Oliver. "'A Drama Unites Them in a Fight to the Death': Some Remarks on the Flourishing of a Cinema of Scientific Vernacularization in France, 1909–1914." *Historical Journal of Film, Radio, and Television*. 22.3 (2002): 353–74. *Taylor Francis Online*. Web. 4 April 2014.

Gaycken, Oliver. "A Secret Life of Plants: Visualizing Vegetative Movement, 1880–1903." *Early Popular Visual Culture*. 10.1 (2012): 51–69. *Taylor Francis Online*. Web. 6 Jan. 2014.

Garis, Howard. "Professor Jonkin's Cannibal Plant." (1905). *Flora Curiosa: Cryptobotany, Mysterios Fungi, Sentient Trees and Deadly Plants in Classic Science Fiction and Fantasy*. Ed. Chad Arment. Landisville, Pa: Coachwhip, 2008. 102–110. Print.

Gassen, H. G. and S. Minol. "Die Alraune Oder Die Sage Vom Galgenmannlein: Science & Fiction." *Biologie in Unserer Zeit*. 36.5 (2006): 302–307. Print.

Glotfelty, Cheryll, and Harold Fromm. *The Ecocriticism Reader: Landmarks in Literary Ecology*. Athens: University of Georgia Press, 1996. Print.

Goethe, Johann Wolfgang. *Der Versuch die Metamorphose der Pflanzen zu erklären*. 1798. *Projekt Gutenberg*. Web. 15 Jan. 2013.

Goodbody, Axel. *The Culture of German Environmentalism: Anxieties, Visions, Realities*. New York: Berghahn Books, 2002. Print.

Goodbody, Axel, and Catherine E. Rigby. *Ecocritical Theory: New European Approaches*. Charlottesville: University of Virginia Press, 2011. Print.

Geheimnisse im Pflanzenleben. Dir. Dr. Ulrich Schulz. Dr. N. Kaufmann, 1931. Film.

George, Stefan. "Komm in den totgesagten Park und Schau." *Sämtliche Werke in 18 Bänden. Band 4: Das Jahr der Seele* Stuttgart: Klett-Cotta, 1982, 12. Print.

Grätzer, Franz. "Paul Scheerbarts letztes Werk" Rev. of *Lesabéndio. Neue metaphysische Rundschau*. (Autumn 1917): 99–101. *Über Paul Scheerbart: 100 Jahre Scheerbart-Rezeption: in Drei Bänden*. Vol. 3. Ed. Michael M. Schardt. Paderborn: Igel, 1992. 223–226. Print.

Griffiths, Alison. "'Shivers down your spine': Panoramas and the Origins of Cinematic Reenactment." *Screen: The Journal of the Society for Education in Film and Television*. 44.1 (2003): 1. *Oxford University Press Journals Online*. Web. 21 Oct. 2013.

Gunning, Tom. "An Aesthetic of Astonishment: Early Film and the [In]Credulous Spectator." *Film Theory and Criticism: Introductory Readings*. Ed. Leo Braudy and Marshall Cohen. New York: Oxford University Press, 1999. 818–32. Print.

Gunning, Tom. "Light, Motion, Cinema!: The Heritage of Loïe Fuller and Germaine Dulac." *Framework: the Journal of Cinema and Media*. 46.1 (2005): 106–129. Print.

Gunning, Tom. "To Scan a Ghost: The Ontology of Mediated Vision." *Grey Room*. 26.4 (2007): 94–127. Print.

Hall, Matthew. *Plants As Persons: A Philosophical Botany*. Albany: State University of New York Press, 2011. Print.

Haeckel, Ernst. *Generelle Morphologie Der Organismen: Allgemeine Grundzüge Der Organischen Formen-Wissenschaft, Mechanisch Begründet Durch Die Von Charles Darwin Reformirte Descendenztheorie*. Berlin: De Gruyter, 1988. Print.

Haeckel, Ernst. *Kunstformen Der Natur*. Leipzig und Wien: Verlag des Bibliographischen Instituts, 1904. Print.

Haraway. Donna, Constance Penley, and Andrew Ross. "Cyborgs at Large: Interview with Donna Haraway." *Social Text*. (1990): 8–23. *Jstor*. Web. 20 Jan. 2014.

Haynes, Roslynn. "From Alchemy to Artificial Intelligence: Stereotypes of the Scientist in Western Literature." *Public Understanding of Science*. 12.3 (2003): 243–253. Print.

Hayungs, Heino. *Die Lehre Von Der Beseeltheit Der Pflanze: Von Fechner Bis Zur Gegenwart*. Leipzig, 1912. Print.

Heidelberger, Michael. *Nature from Within: Gustav Theodor Fechner and His Psychophysical Worldview*. Pittsburgh: University of Pittsburgh Press, 2004. Print.

Heise, Ursula K. *Sense of Place and Sense of Planet: The Environmental Imagination of the Global.* Oxford: Oxford University Press, 2008. Print.

Helfer, Martha B. "Gender Studies and Romanticism." *The Literature of German Romanticism.* Vol. 8. Camden House, 2004. 229–50. Print.

Hermand, Jost. *Grüne Utopien in Deutschland: Zur Geschichte Des Ökologischen Bewusstseins.* Frankfurt am Main: Fischer Taschenbuch, 1991. Print.

Hoffmann, E. T. A. *Der Sandmann. Projekt Gutenberg.* Web. 23 Aug. 2012.

Hofmannsthal, Hugo von. "Die Rose und Der Schreibtisch." *Gesammelte Werke in 10 Einzelbänden. Erzählungen, erfundene Gespräche und Briefe, Reisen.* Frankfurt am Main: Fischer, 1979. 443. Print.

Huysmen, J. K. *A rebours.* Paris: Ferroud, 1920. Print.

Hofmannsthal, Hugo. *Ein Brief.* Darmstadt: Ernst Ludwig Presse, 1925. Print.

Hofmannstal, Hugo. "Die Rose und der Schreibtisch." *Gesammelte Werke in 10 Einzelbänden. Erzählungen, erfundene Gespräche und Briefe, Reisen.* Frankfurt am Main: Fischer, 1979. 443. Print.

Hormonwirkungen bei höheren Pflanzen. Dir. Prof. Dr. Kurt Noack. Aus dem Pflanzenphysiologischen Institut der Universitaet Berlin. 1920–29. *Filmarchiv-Bundesarchiv.* Film.

Hormonwirkungen bei höheren Pflanzen. Dir. Prof. Dr. Kurt Noack. Aus dem Pflanzenphysiologischen Institut der Universitaet Berlin. 1920–29. Censorship card. *Filmarchiv-Bundesarchiv.* Microfilm.

Hyazinthe, Daisy Spieß. Das Blumenwunder Programm. 1926. *Art.tv.de.* Web. 5 Jan. 2014.

Insektenfressende Pflanzen. Dir. Georg E. F. Schulz. Ufa-Unterrichts-Film, 1929. Censorship card. *Filmarchiv-Bundesarchiv.* Microfilm.

Jackson, Tony E. "Writing and the Disembodiment of Language." *Philosophy and Literature.* 27.1 (2003): 116–133. Print.

Kaes, Anton. "Film in der Weimarer Republik." *Geschichte Des Deutschen Films.* Ed. Wolfgang Jacobsen, Anton Kaes, and Hans H. Prinzler. Stuttgart: Metzler, 2004. 39–99. Print.

———. *Shell Shock Cinema: Weimar Culture and the Wounds of War.* Princeton: Princeton University Press, 2009. Print.

Kelley, Theresa M. *Clandestine Marriage: Botany and Romantic Culture.* Baltimore: The John Hopkins University Press, 2012. *Ebrary.* Web.

Killen, Andreas. *Berlin Electropolis: Shock, Nerves, and German Modernity.* Berkeley: University of California Press, 2006. Web.

Kluittenberg, Eric. "On the Archaeology of Imaginary Media." *Media Archaeology: Approaches, Applications, and Implications.* Berkeley: University of California P, 2011. 48–69. Print.

Kracauer, Siegfried. *From Caligari to Hitler: A Psychological History of the German Film.* Ed. Leonardo Quaresima. Princeton, N.J.: Princeton University Press, 2004. Print.

Kuzniar, Alice. "A Higher Language: Novalis on Communion with Animals." *The German Quarterly*. 76.4 (2003): 426–442. *Jstor*. Web. 1 Dec. 2013.

Laban, Rudolf von. *Gymnastik und Tanz*. Oldenburg: Verlag Gerhard Stalling, 1925. Print.

Lachman, Gary. *A Dark Muse: A History of the Occult*. New York: Thunder's Mouth Press, 2005. Print.

Lafitte. *The Dance of the Lily. Fifteen Years of a Dancer's Life: With Some Account of Her Distinguished Friends*. By Loië Fuller. London: H. Jenkins Ltd, 1913. 93. *Internet Archive*. Web. 27 Sept. 2013.

Lambert, Carrie. "On Being Moved: Rainer and the Aesthetics of Empathy," *Yvonne Rainer: Radical Juxtapositions 1961–2002*. Ed. Sid Sachs. Philadelphia: The University of the Arts, 2003. 45–6. Print.

Lamothe, Kimerer L. *Nietzsche's Dancers: Isadora Duncan, Martha Graham, and the Revaluation of Christian Values*. New York: Palgrave, 2006. Print.

Laist, Randy. *Plants and Literature: Essays in Critical Plant Studies*. Amsterdam: Rodopi, 2013. Print.

Laßt Blumen Sprechen (*Quand Les Fleurs Parlent*). 1929. *Filmarchiv-Bundesarchiv*. Film.

Lasswitz, Kurd. *Atomistik und Kriticismus: Ein Beitrag zur erkenntnistheoretischen Grundlegung der Physik*. Braunscheig: Vieweg, 1878. *Internet Archive*. Web. 6 Jan. 2015.

Lasswitz, Kurd. *Auf zwei Planeten: Roman in zwei Büchern*. Weimar: Felber, 1897. Print.

Lasswitz, Kurd. "Die enflohene Blume: Eine Geschichte vom Mars." *Empfundenes und Erkanntes: aus dem Nachlasse*. Leipzig: Elischer, 1919. 174–86. *Internet Archive*. Web. 5 Dec. 2014.

Lasswitz, Kurd. *Die Lehre Kants von der Idealität des Raumes und der Zeit*. Berlin: Weidmann, 1883. *Internet Archive*. Web. 15 Jan. 2015.

Lasswitz, Kurd. "Einleitung des Herausgebers." *Nanna, Oder, Über Das Seelenleben Der Pflanzen*. 4th Ed. Leipzig: L. Voss, 1908. III–IX. *Internet Archive*. Web. 17 Sept. 2013.

Lasswitz, Kurd. Introduction. *Nanna, oder über das Seelenleben der Pflanzen*. By Gustav Fechner. Leipzig: Voss, 1908. III–IX. *Internet Archive*. Web. 30 Oct. 2015.

Lasswitz, Kurd. *Geschichte der Atomistik vom Mittelalter bis Newton*. Leipzig: Voss, 1890. *Internet Archive*. Web. 30 Nov. 2015.

Lasswitz, Kurd. "Naturnothwendigkeit und ihre Grenzen." *Nord und Süd*. Jan.–Feb.–March 1893. *Internet Archive*. Web. 28 Dec. 2014.

Lasswitz, Kurd. *Schlangenmoos und Sternentau: Ungekürzte Sonderausgabe der erstmals zwischen 1884 und 1909 erschienenen Romane Schlangenmoos, Homchen, Aspira, und Sternentau*. Lüneberg: Dieter von Reeken, 2012. Print.

Lasswitz, Kurd. *Sternentau: Die Pflanze Vom Neptunsmond*. Leipzig: Elischer, 1909. Print.

Lasswitz, Kurd. "Über Zukunftsträume." *Wirklichkeiten: Beiträge zum Weltverständnis*. Leipzig: L Elischer Nachfolger, 1903. 423–45. Print.

Lasswitz, Kurd. "Unser Recht auf Bewohner andrer Planeten." *Empfundenes und Erkanntes: aus dem Nachlasse*. Leipzig: Elischer, 1919. 174–86. *Internet Archive*. Web. 5 Dec. 2014.

Lovejoy, Arthur O. *The Great Chain of Being: A Study of the History of an Idea*. Cambridge: Harvard University Press, 1961. Print.

M-1, Dr. "Kultur- oder Lehrfilm? Kritische Betrachtungen zum 'Blumenwunder'." *Der Lehrfilm: Beilege zu "Der Filmspiegel."* Kinematographische Monatsheft, Berlin, July 1926. *Archiv der deutschen Kinemathek*. Print.

Maeterlinck, Maurice. "Scents." *The Intelligence of Flowers*. Albany: State University of New York Press, 2008. 63–71. Print.

Maeterlinck, Maurice. "The Intelligence of Flowers." *The Intelligence of Flowers*. Albany: State University of New York Press, 2008. 1–61. Print.

Maeterlinck, Maurice. *The Treasure of the Humble*. New York: Dodd, Mead & Co, 1899. Print.

Makartstrauß. 1884. *Scherl's Magazin*. 6 June 1933: 340. *Illustrierte Magazine*. Web. 10 Jan. 2014.

Marder, Michael. *Plant-Thinking: A Philosophy of Vegetal Life*. New York: Colombia University Press, 2013. Print.

Marks, Jonathan. "Great Chain of Being" *Encyclopedia of Race and Racism*. Ed. John H. Moore. Vol. 2. Detroit, Mich: Macmillan Reference USA, 2008. 68–73. *Oxford Reference Online*. Web. 31 July 2013.

Marks, Laura U. "Video Haptics and Erotics." *Screen: the Journal of the Society for Education in Film and Television*. 39.4 (1998): 331. *Oxford University Press Journals Online*. Web. 15 Nov. 2013.

Martin, John. *The Modern Dance*. (1933). Princeton: Princeton Book, 1989. Print.

Martin, John. "Dance as a Means of Communication." *What is Dance?*. Ed. Roger Copeland and Marshall Cohen (Oxford: Oxford University Press, 1983), 22. Print.

Mayne, Judith. "Dracula in the Twilight: Murnau's *Nosferatu* (1922)." *German Film & Literature: Adaptations and Transformations*. Ed. Eric Rentschler. New York: Methuen, 1986. 25–39. Print.

McCormick, Richard W. *Gender and Sexuality in Weimar Modernity: Film, Literature, and "New Objectivity."* New York: Palgrave, 2001. Print.

Merkel, Friedrich. *Das Mikrosop und seine Anwendung*. München: R. Oldenbourg, 1875. *GoogleBooks*. Web. 18 Jan. 2015.

Metropolis. Dir. Fritz Lang. Perf. Brigitte Helm. New York, NY: Kino on Video, 2002. Film.

Meyrink, Gustav. "Bologneser Tränen." *Orchideen: Sonderbare Geschichten*. Albert Langen: München, 1905. 77–84. *Internet Archive*. Web. 24 Jan. 2011.

Meyrink, Gustav. "Der Kardinal Napellus." *Fledermäuse: Sieben Geschichten*. Kurt Wolff: Leipzig, 1916. 177–199. Print.

Meyrink, Gustav. "Die Pflanzen des Doktor Cinderella." *Wachsfigurenkabinett: Sonderbare Geschichten*. München: Albert Langen, 1908. 185–203. *Internet Archive*. Web. 17 Jan. 2013.

Meyrink, Gustav. "Die Verwandlung des Blutes." *Fledermäuse: Erzählungen, Fragmente, Aufsätze*. Frankfurt am Main: Ullstein, 1992. Print.

Miller, Elaine. *The Vegetative Soul: From Philosophy of Nature to Subjectivity in the Feminine*. Albany: State University of New York Press, 2002. *Google Play*. Web. 20 Nov. 2013.

Milton, Kay. *Loving Nature: Towards an Ecology of Emotion*. London: Routledge, 2002. *Ebrary*. Web. 9 April 2014.

Monaco, Paul. *Cinema and Society: France and Germany During the Twenties*. New York: Elsevier, 1976. Print.

Müller, Hedwig and Patricia Stöckemann: *... jeder Mensch ist ein Tänzer: Ausdruckstanz in Deutschland zwischen 1900 und 1945*. Anabas: Gießen, Begleitbuch azur Ausstellung: "Weltenfriede—Jugendglück" Vom Ausdruckstanz zum Olympischen Festspiel 1939. Print.

Murnau, F. W. *Nosferatu: A Symphony of Horror*. Chatsworth, CA: Image Entertainment, 2000. Film.

Natur und Liebe—Vom Untier zum Menschen. Dir. Ulrich Schulz. Berlin: Ufa, 1927. *Filmarchiv-Bundesarchiv*. Film.

Nee, Sean. "The Great Chain of Being." *Nature*. 435.7041 (2005): 429. Print.

Nierendorf, Karl. Introduction. *Urformen*. 1928. *Karl Blossfeldt: The Complete Published Work*. By Karl Blossfeldt. Ed. Hans-Christian Adam. Köln: Taschen, 2008. 25–29. Print.

Nietzsche, Friedrich. *The Birth of Tragedy and Other Writings*. Ed. Raymond Geuss and Ronald Speirs. Trans. Ronald Speirs. Cambridge: Cambridge UP, 1999. Print.

Nietzsche, Friedrich W., Maudemarie Clark, and Brian Leiter. *Daybreak: Thoughts on the Prejudices of Morality*. Cambridge, U.K: Cambridge University Press, 1997.

Nosferatu. Dir. F. W. Murnau. Minneapolis, Minn.: Mill Creek Entertainment, LLC, 2006. Film.

Paleologue, Jean de. *La Loïe Fuller. Folies-Bergè. Art.com*. Web. 27 Jan. 2014.

Parikka, Jussi. *Insect Media: An Archaeology of Animals and Technology*. Minneapolis: University of Minnesota Press, 2010. *Ebrary*. Web. 11 Dec. 2013.

Partsch, Cornelius. "Paul Scheerbart and the Art of Science Fiction." *Science Fiction Studies*. 29.2 (2002): 202–220. Print.

Pfankuch, Kai. "Die Weltfluchten des Paul Scheerbart" (1986) *Über Paul Scheerbart: 100 Jahre Scheerbart-Rezeption: in Drei Bänden*. Vol. 1. Ed. Berni Lörwald. Paderborn: Igel, 1992. 137–49. Print.

Pfeffer, Wilhelm. *Pflanzenphysiologie: Ein Handbuch Der Lehre Vom Stoffwechsel Und Kraftwechsel in Der Pflanze*. Leipzig, 1904. Print.

Pflanzen leben. Dir. Hubert Schonger. Naturfilm, 192?. Film.

Pollan, Michael. *The Botany of Desire: A Plant's Eye View of the World*. New York: Random House, 2001. Print.

Qasim, Mohammad. *Gustav Meyrink: Eine Monographische Untersuchung*. Stuttgart: Heinz, 1981. Print.

Raabe, Paul. Afterword. "Lesabéndio." (1964). *Über Paul Scheerbart: 100 Jahre Scheerbart-Rezeption: in Drei Bänden*. Vol. 1. Ed. Berni Lörwald. Paderborn: Igel, 1992. 54–61. Print.

Rapine, M. *Electric Discharge in Rarified Gases. Elementary Treatise on Natural Philosophy, Part 3 Electricity and Magnetism*. (1869) By Augustin Privat Deschanel. Trans. Joseph David Everett. 13th ed. New York: Appleton & Co., 1896. 569. *Google books*. Web. 27 Jan. 2014.

Räuber in der Natur. (Fleischfressende Pflanzen). Emelka-Film, 1930. Censorship card. *Filmarchiv-Bundesarchiv*. Microfilm.

Rausch, Mechthild. "70 Trillionen Weltgrüsse." *70 Trillionen Weltgrüsse: Eine Biographie in Briefen 1889–1915*. Berlin: Argon, 1991. 612–637. Print.

Rathunde, K. "Montessori and Embodied Education." *Namta Journal*. 33 (2008): 187–216. Print.

Renger-Patzsch, Albert. *Inneres einer Orchideenblüte. Der Querschnitt*. 6 March 1926: 41. *Illustrierte Magazine*. Web. 12 Aug. 2013.

Renger-Patzsch, Albert. *Orchideenblüte (Brassia verrucosa). Querschnitt*. October 1925: 56. *Illustrierte Magazine*. Web. 10 January 2014.

Richards, Robert J. *The Romantic Conception of Life: Science and Philosophy in the Age of Goethe*. Chicago: University of Chicago Press, 2002. Print.

Richter, Hans. *Der Zweigroschenzauber*. 1928. Film.

Riebicke, Gerhard. *Tanz im Freien. Querschnitt*. October 1925: 57. *Illustrierte Magazine*. Web. 10 Jan. 2014.

Rolli, Beatrice. *Paul Scheerbarts (weltgestaltende Phantasiekraft) Zwischen Utopie Und Phantasmagorie: Eine Interpretation von Münchhausen Und Clarissa: Ein Berliner Roman" Als Einführung Ins Gesamtwerk*. Zürich: Theater am Neumarkt, 1983. Print.

Rottensteiner, Franz. "Ordnungsliebend im Weltraum." in *Polaris 1*. Frankfurt am Main: Insel, 1973. 133–4. Print.

Ruosch, Christian. *Die Phantastisch-Surreale Welt Im Werke Paul Scheerbarts*. Bern: Herbert Lang, 1970. Print.

Ruttmann, Walter. *Das wiedergefundene Paradies*. 1925. Film.

Sam, der Königstiger im Londoner Zoo. London. *Der Querschnitt*. 6 March 1926: 40. *Illustrierte Magazine*. Web. 12 Aug. 2013.

Sandford, John. "Chaos and Control in the Weimar Film." *German Life and Letters*. 48.3 (1995): 311. Print.

Schorske, Carl. *Fin de Siècle Vienna: Politics and Culture*. Cambridge: Cambridge University Press, 1981. 279–322. Print.

Scheerbart, Paul. "[Autobiographie] (6.7.1904)" *Verein für Kunst. Winterprogram 1904/05*. *Gesammelte Werke*. Ed. Thomas Bürk, Joachim Körber, and Uli Kohnle. Linkenheim: Phantasia, 1986. 20–21. Print.

Scheerbart, Paul. "Das Ende des Individualismus. Eine Kosmologische Betrachtung." *Die Gesellschaft*. August 1895. *Gesammelte Werke*. Ed. Thomas Bürk, Joachim Körber, and Uli Kohnle. Linkenheim: Phantasia, 1986. 1093–1097. Print.

Scheerbart, Paul. *Das Große Licht: Ein Münchhausen Brevier. Projekt Gutenberg*. Web. 12 June 2011.

Scheerbart, Paul. "Die Ästhethik der Phantastik." *Amsler & Ruthard's Wochenberichte 1894/95*. *Gesammelte Werke*. Ed. Thomas Bürk, Joachim Körber, and Uli Kohnle. Linkenheim: Phantasia, 1986. 2–3. Print.

Scheerbart, Paul. "Die Entwicklung des Luftmilitarismus und die Auflösung der europäischen Land-Heere, Festungen und Seeflotten." *Gesammelte Werke*. Ed. Thomas Bürk, Joachim Körber, and Uli Kohnle. Linkenheim: Phantasia, 1986. Print.

Scheerbart, Paul. *Die Grosse Revolution. Ein Mondroman*. S.l.: Tradition Classics, 2012. Print.

Scheerbart, Paul. "Die Phantastik im Kunstgewerbe." *Das Atelier*. 1890/91. *Gesammelte Werke*. Ed. Thomas Bürk, Joachim Körber, and Uli Kohnle. Linkenheim: Phantasia, 1986. 4–5. Print.

Scheerbart, Paul. "Die Phantastik in der Malerei." *Freie Bühne für modernes Leben* 1891. *Gesammelte Werke*. Ed. Thomas Bürk, Joachim Körber, and Uli Kohnle. Linkenheim: Phantasia, 1986. 286–290. Print.

Scheerbart, Paul. "Flora Mohr: Eine Glasblumen-Novelle." *Dichterische Hauptwerke*. Stuttgart: Goverts, 1962. 489–519. Print.

Scheerbart, Paul. *Glasarchitektur*. München: Rogner & Bernhard, 1971. Print.

Scheerbart, Paul. "Licht und Luft." *Ver Sacrum* 7 July 1898. *Gesammelte Werke*. Ed. Thomas Bürk, Joachim Körber, and Uli Kohnle. Linkenheim: Phantasia, 1986. 13–14. Print.

Scheerbart, Paul. *Liwûna Und Kaidôh: Ein Seelenroman*. Frankfurt am Main: Suhrkamp, 1990. Print.

Scheerbart, Paul. *Munchhausen und Clarissa*. S.l.: Paderborner Grossdruckbuch, 2013.

Scheerbart, Paul. "Münchhausen und Klarissa [Selbstanzeige]" *Die Zukunft* 1905/06. *Gesammelte Werke*. Ed. Thomas Bürk, Joachim Körber, and Uli Kohnle. Linkenheim: Phantasia, 1986. 462–63. Print.

Scheerbart, Paul. *Münchhausen und Clarissa: ein Berliner Roman. Projekt Gutenberg*. Web. 5 Apr. 2011.

Scheerbart, Paul. "To Richard Dehmel." 11 Dec. 1896. *70 Trillionen Weltgrüsse: Eine Biographie in Briefen 1889–1915*. Ed. Mechthild Rausch. Berlin: Argon, 1991. 36. Print.

Scheerbart, Paul. "Rahmenkunst" *Das neue Jahrhundert* 1898/99. *Gesammelte Werke*. Ed. Thomas Bürk, Joachim Körber, and Uli Kohnle. Linkenheim: Phantasia, 1986. 486–492. Print.

Scheerbart, Paul. "Sind die Kometen lebendige Wesen?" *Die Gegenwart*. 1910. *Gesammelte Werke*. Ed. Thomas Bürk, Joachim Körber, and Uli Kohnle. Linkenheim: Phantasia, 1986. 257–59. Print.

Scheerbart, Paul. "Sternschnuppen und Kometen." *Die Gegenwart*. 1909. *Gesammelte Werke*. Ed. Thomas Bürk, Joachim Körber, and Uli Kohnle. Linkenheim: Phantasia, 1986. 242–44. Print.

Schelling, Friedrich W. J. *Einleitung Zu Seinem Entwurf Eines Systems Der Naturphilosophie: Oder; Ueber Den Begriff Der Speculativen Physick Und Die Inere Organisation Eines Systems Dieser Wissenschaft*. Leipzig: C. E. Gabler, 1799. Print.

Schiller, Friedrich. *Der Geisterseher: Aus den Memoiren des Grafen von O****. Leipzig: s.n., 1810. Web. 14 Jan. 2015.

Schiller, Friedrich. "Philosophie of Physiologie." *Medizinische Schriften*. Miesbach: Mayr, 1959.

Schlegel. Friedrich. *Fr. Schlegels Lucinde*. Hamburg: Schubert & Comp, 1842. *Internet Archive*. Web. 10 April 2014.

Schmidlin, Eduard. *Populäre Botanik oder gemeinfassliche Anleitung zum Studium der Pflanze und des Pflanzenreiches*. 2nd ed. Stuttgart: Gustav Weise, 1867. *Google Books*. Web. 12 Dec. 2013.

Section of the Rotunda, Leicester Square, In Which is Exhibited the Panorama. Plans, and Views in Perspective, with Descriptions, of Buildings Erected in England and Scotland: And Also an Essay, to Elucidate the Grecian, Roman and Gothic Architecture, Accompanied with Designs. By Robert Mitchell. London: Printed by Wilson & Co. for the author, 1801. *Google search*. Web. 28 Jan. 2014.

Seitler, Heino. "Kriminalität im Zirkus." *Das Kriminal-Magazin*. 3 August 1931, 1667–70. *Illustrierte Magazine*. Web. 12 Aug. 2013.

Serres, Michel. *The Natural Contract*. Ann Arbor: University of Michigan Press, 1995. Print.

Shelley, Mary W., and J. P. Hunter. *Frankenstein: The 1818 Text, Contexts, Nineteenth-Century Responses, Modern Criticism*. New York: W. W. Norton, 1996. Print.

Shepherd, V. A. "At the Roots of Plant Neurobiology: A Brief History of the Biophysical Research of J. C. Bose." *Science and Culture*. 78: 5–6 (2012): 196–210. Web. *ArticleFirst*. 17 Nov. 2014.

Shields, Christopher. "Aristotle's Psychology." *Stanford Encyclopedia of Knowledge*. Ed. Edward N. Zalta. 2010. *Stanford Encyclopedia of Knowledge*. Web. 20 Aug. 2013.

Sicks, Kai. "Der Querschnitt: oder die Kunst des Sportreibens." *Leibhaftige Moderne: Körper in Kunst Und Massenmedien 1918 Bis 1933*. Ed. Michael Cowan, and Kai M. Sicks. Bielefeld: transcript, 2005. 33–47. Print.

Smyth, Mary M. "Kinesthetic Communication in Dance." *Dance Research Journal*. 16: 2 (Autumn, 1984): 19–22, Congress on Research in Dance. *Jstor*. Web. 7 Nov. 2011.

Sobchack, Vivian C. *Carnal Thoughts: Embodiment and Moving Image Culture*. Berkeley: University of California Press, 2004. *Ebrary*. Web. 20 Nov. 2013.

Steiner, Uwe. *Walter Benjamin: An Introduction to his Work and Thought*. Trans. Michael Winkler. Chicago: U of Chicago Press, 2010. Print.

Sternberger, Dolf and Joachim Neugroschel. "Panorama of the 19th Century" *October*. 4. (Autumn 1977): 3–20. *Jstor*. Web. 11 Aug. 2012.

Stoker, Bram, Nina Auerbach, and David J. Skal. *Dracula: Authoritative Text, Contexts, Reviews and Reactions, Dramatic and Film Variations, Criticism*. New York: W. W. Norton, 1997. Print.

Studlar, Gaylyn. "Masochism and the Perverse Pleasures of the Cinema." *Film Theory and Criticism: Introductory Readings*. 4th Ed. ed. Gerald Mast, Marshall Cohen and Leo Braudy. New York: Oxford UP, 1992. 773–790.

Sweeny, Marvin. *The Oxford Dictionary of the Jewish Religion*. Ed. Adele Berlin and Maxine Grossman. New York: Oxford University Press, 2011. Print.

"Tägliche Rundschau" *Presse über den Film "Das Blumenwunder". Archiv der deutschen Kinemathek*.

Tresch, John. *The Romantic Machine: Utopian Science and Technology After Napoleon*. Chicago: U of Chicago P, 2012. Print.

von Uexküll, Jakob. "The Theory of Meaning" (1940). *Essential Readings in Biosemiotics: Anthology and Commentary*. 90–114. Print.

Verdeja, Ernesto. "Adorno's Mimesis and its Limitations for Critical Social Thought." *European Journal of Political Theory*. (2009) 8: 493. Web. 17 Aug. 2011.

Waxworks. New York, NY: Kino on Video, 2002. Film.

Weber, Max. "Wissenschaft als Beruf"(1919) *Gesammelte Aufsaetze zur Wissenschaftslehre*. Internet Archive. 524–555.

Wege zur Kraft und Schönheit. Dir. Wilhelm Prager and Nicholas Kaufmann, UfA, 1924, 1925. Film.

Weinstein, Valerie. "Alraune: The Vamp and the Root of Horror." *The Many Faces of Weimar Cinema: Rediscovering Germany's Filmic Legacy*. Ed. Christian Rogowski. Rochester, N.Y: Camden House, 2010. 198–210. Print.

Wells, H. G. *The Island of Doctor Moreau*. New York: Gardening City Publishing, 1896. Print.

Wiesenthal, Grete, perf. *Den okända*. Dir. Mauritz Stiller. Svenska Biografteatern AB, 1913. *Das fremde Mädchen*. By Hugo von Hofmannsthal. Film.

Wilde, Ann and Jürgen Wilde, eds. Karl Blossfeldt: *Working Collages*. Cambridge, Mass.: MIT Press, 2001. 1–21. Print.

Worringer, Wilhelm. *Abstraktion und Einfühlung: ein Beitrag zur Stilpsychologie*. Munich: R Piper, 1911.

Wunder der Pflanzenwelt. Kricheldorff-Lehrfilm, 1925. Film.

Wunder der Natur: Aus den Wurzeln kommt der Kraft. 1925–35. Censorship card. *Filmarchiv-Bundesarchiv*. Microfilm.

Index

Printed in the United States
By Bookmasters